SCIENCING

An Involvement Approach
to Elementary Science Methods

Sandra E. Cain
and
Jack M. Evans

CENTRAL MICHIGAN UNIVERSITY

Charles E. Merrill Publishing Company
A Bell & Howell Company
Columbus   Toronto   London   Sydney

Published by
Charles E. Merrill Publishing Company
A Bell and Howell Company
Columbus, Ohio 43216

This book was set in Times Roman and Megaron.

The cover was prepared by Will Chenoweth.

Library of Congress Catalog Card Number: 78-71088
International Standard Book Number: 0-675-08364-8

Printed in the United States of America
1 2 3 4 5 6 7/ 85 84 83 82 81 80 79

CONTENTS

PREFACE

This program is designed to offer elementary science methods instructors, preservice teachers, and inservice teachers an exciting and challenging alternative to the traditional lecture-discussion format prevalent in most college classes. Direct involvement is utilized throughout to guide students in acquiring the specific competencies necessary to teaching science in the elementary school. The program is structured to provide flexibility, interaction, individualization and personalization.

Six broad Units are included. Each Unit is composed of one or more parts that focus on identified competency areas. Each of these Parts contains an introduction, overall goals, a cluster of specific performance behaviors which the teacher is to acquire, a background reading section, and a list of supplementary reading material. The core of each Part, however, is a set of activities arranged and designed to guide the students toward the acquisition of the performance behaviors. Space is provided for recording and organizing information gathered during the activity portions.

This program offers preservice and inservice teachers a chance to become directly involved in sciencing. Students are exposed to science education philosophy, direct first-hand experience involving science equipment and materials, and several science concepts. Generally, the program encourages students to be active participants in the learning process. Representative commercial programs are included to acquaint students with the laboratory, textbook, and text-kit approaches to sciencing. No particular program is advocated, and hopefully the cross application of concepts, skills, and methods between various commercial programs and traditional materials will be evident.

This is not a programmed textbook. It is based on the philosophy of student-instructor involvement. The students read informational materials, try hands-on as well as mental activities, discuss reactions with peers and instructor, record information gathered, and apply what they learn to an evaluation task (or exam, if you prefer). The instructor initiates the Units and each Part (with an introductory presentation to individuals, small groups, or large groups), confers with students during the course of activities, modifying or prescribing as necessary, and participates in summarizing the learning experience upon completion. The instructor also evaluates the students' performances on the evaluation task and prescribes remediation if warranted. *The student and instructor are partners in a learning experience, actively working together to analyze individual needs and utilize the program to meet those needs.* Some students will not need all portions of the program; such portions may be modified or adapted.

ACKNOWLEDGMENTS

The authors wish to thank their students at Central Michigan University who contributed their time and energy to provide valuable suggestions during the preparation of this material. We also appreciate the help of the local Mount Pleasant elementary teachers who were involved during the beginning stages in identifying specific areas of competency. We also want to recognize Dr. Curtis Nash, our dean, and Dr. Robert Oana, our department chairman, for their engagement and understanding during this time consuming task. A special thanks is offered to Jan Hansen, our typist. She demonstrated a fantastic amount of patience throughout the entire project. Dr. Cain offers her appreciation to Dr. Anita Bozardt for her contribution to the Classroom Management Unit. Thanks are also extended to our photographers, Brenda Hunt, David Brittain, and Flint Horton for the excellent photographs included in the text.

Last, but certainly not least, we want to offer our thanks to our families for their understanding and tolerance of the many long hours that were necessary to complete this project.

Sandra E. Cain
Jack M. Evans

PHOTO CREDITS

SCIENCING

An Involvement Approach
to Elementary Science Methods

UNIT 1

Process-oriented Science

In the past, science has been approached as a body of knowledge, or facts, to be memorized and repeated later on a test. The 1960s saw a movement in science away from the product, or content, emphasis toward a *process* orientation. Science was becoming more of a "doing" thing. Science educators began using the term *sciencing* to focus on this change of approach. The hands-on, process-oriented kit approach to elementary science was introduced into many elementary schools. These new curriculum projects were among the most exciting things that had ever happened to elementary science.

Support from the National Science Foundation and the U.S. Office of Education made it possible for scientists, science educators, teachers, and children to be directly involved in the development of these process-oriented programs. The research and theories of well-known child development psychologists, Jean Piaget, Robert Gagné, and Jerome Bruner, provided much direction and guidance for the new science approach.

Consequently, heavy reliance on the textbook and the teacher for information has given way to a more direct hands-on approach. Science or sciencing is beginning to be seen as a *means* rather than the end product. Elementary children are involved in generating, organizing, and evaluating the content, not merely in memorizing it. In order to be successful with the new approach, the learner must develop certain process-inquiry skills. The following basic skills are considered appropriate for grades K-3:

1. *Observing*—using the senses to find out about subjects and events
2. *Classifying*—grouping things according to similarities or differences
3. *Measuring*—making quantitative observations
4. *Using spatial relationships*—identifying shapes and movement
5. *Communicating*—using the written and spoken word, graphs, drawings, diagrams, or tables to transmit information and ideas to others
6. *Predicting*—making forecasts of future events or conditions based upon observations or inferences
7. *Inferring*—explaining an observation or set of observations

The following integrated skills are appropriate for grades 4-8:

1. *Defining operationally*—creating a definition by describing what is done and observed
2. *Formulating hypotheses*—making educated guesses based on evidence that can be tested
3. *Interpreting data*—finding patterns among sets of data which lead to the construction of inferences, predictions, or hypotheses
4. *Controlling variables*—identifying the variables of a system and selecting from the variables those which are to be held constant and those which are to be manipulated in order to carry out a proposed investigation
5. *Experimenting*—investigating, manipulating, and testing to determine a result

Process-inquiry skills are not separate from science content. Rather, they are the "tools" of scientific investigation. The utilization of these skills in gathering, organizing, analyzing, and evaluating science content is an on-going goal of sciencing.

PART 1-1

Sciencing

FLOWCHART: Sciencing

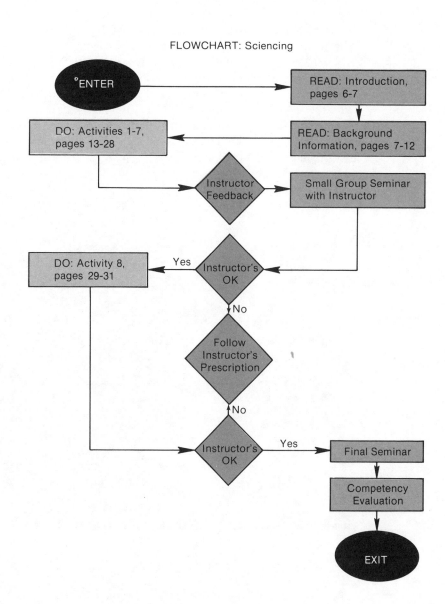

Introduction

How do teachers decide what is worth teaching? What criteria should be used for making decisions about what students will learn in science? These are questions that concerned teachers and preservice teachers ask. Unfortunately, they are very complex questions with no easy answers.

A logical way to seek possible solutions to these questions might be to begin by examining the learning process itself. The background reading of this Part contains information that should provide insight into the nature of learning. This information should establish a rationale for the new approach of process-oriented science, or sciencing. Part 1-1 is designed to help mesh theory with practice by involving you with ideas, materials, and activities that will help you acquire specific skills necessary for achieving competence in the area of sciencing.

Process-inquiry skills are basic to all later learning. The laboratory activities included in this Part deal with the basic process-inquiry skills of observing, using spatial relationships, measuring, classifying, communicating, predicting, inferring, and the integrated process skills of defining operationally, formulating hypotheses, interpreting data, controlling variables, and experimenting. In doing the activities, you will become aware of and utilize these skills in gathering, organizing, analyzing, and evaluating data. In some of the activities, you will be involved in small group work, and in others, you will work alone. Your instructor will serve as a resource person.

A flowchart is provided with each Part. Use them as road maps to tell you where you are going and how to get there. You will want to refer to them frequently.

These students are using process-inquiry skills in a laboratory experiment.

GOALS

After completing this Part, you will demonstrate competency in the following:
A. Assimilating and accommodating Piaget's theory of mental development
B. Acquiring the process skills

BEHAVIORAL OBJECTIVES

In completing this Part, you will do the following:
A. Identify and describe Piaget's four stages of mental development
B. Explain how Piaget's theory of mental development has influenced elementary science curriculum
C. Identify and demonstrate the ability to utilize the following process-inquiry skills:
 1. Observing
 2. Classifying
 3. Measuring
 4. Using spatial relations
 5. Communicating
 6. Predicting
 7. Inferring
 8. Integrated processes (defining operationally, formulating hypotheses, interpreting data, controlling variables, and experimenting)

Background Information

The new approach to science has caused a change in the role of both the elementary teacher and the elementary student. No longer is the teacher's main role that of information giver. No longer is the elementary student viewed as a sponge to "soak up" the information given. The new approach demands *active* participation on the part of the student with the teacher serving as a guide and resource person. With this approach, learning experiences foster growth and development in all areas of learning, not just in the memorization of facts.

Domains of Learning

Generally, there are three areas, or domains, of learning that an elementary teacher must consider. They are cognitive, psychomotor, and affective.[1] All learning can be classified according to these three domains. Let's look at each area more closely.

COGNITIVE

The cognitive domain encompasses all learning concerned with the acquisition of facts, concepts, and generalizations. There are two methods for obtaining knowledge. One method involves gaining knowledge almost solely from reading, listening to lectures, and other secondary sources. This approach tends to emphasize the content, or product, aspect of science. The second method of obtaining knowledge is through direct, firsthand experience. This approach utilizes empirical procedures involving process-inquiry skills.

---

[1]For more information on these three domains, see the books by Bloom, et al., Krathwohl et al., and Harrow in the Suggested Reading list.

These children are learning by combining touching and questioning.

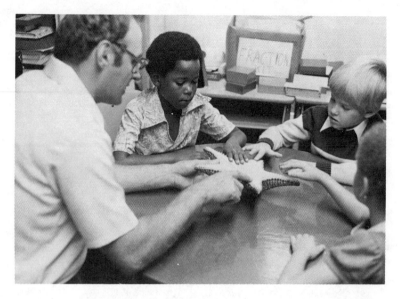

Most science educators agree that both approaches are necessary for effective learning to take place. Secondary sources are legitimate sources of scientific information. But they do differ from direct first-hand manipulation of equipment and materials.

PSYCHOMOTOR

The psychomotor domain deals with physical skills. In science, we are concerned with providing opportunities for elementary children to manipulate equipment and materials in performing laboratory tasks. They are expected to be able to manipulate laboratory equipment with some facility so that they do not hurt themselves or others and do not damage the equipment.

AFFECTIVE

The affective domain involves attitudes, interests, and feelings. Most science teachers hope that their students will like science. In addition, it is hoped that students will develop and accept certain attitudes associated with scientific inquiry. The acceptance and utilization of the processes of scientific inquiry—observing, measuring, hypothesizing, formulating generalizations, and designing and conducting experiments—by elementary children is considered an important goal of science instruction.

As an elementary science teacher, you will be responsible for providing learning experiences for children which will help them grow and develop in each of the three domains of learning.

Stages of Development

As stated earlier, science curricula have been influenced by the research of many noted child development theorists. Jean Piaget, a Swiss psychologist, has been a major influence in promoting the theory that the ability to think and learn is itself a growing thing. Educators who accept this theory see their roles as that of providing learning experiences to help children develop their mental capacity.

Piaget's research in the area of mental development in children has resulted in four identified stages of mental development:

1. Sensorimotor
2. Preoperational
3. Concrete operations
4. Formal or abstract thought [2]

These stages are sequential. Each child passes through each stage in the same order, but not necessarily at the same rate. The attainment of formal, or abstract, thought—the highest level of development—is not achieved by most children until around the age of eleven or twelve. But in order to reach this final stage, children must be provided with opportunities to develop the prerequisite skills of the proceeding stages. You can begin to see how important curricular decisions—what to teach, how to teach, when to teach—are affected by a child's particular stage of development. The science curriculum is a vital and important part of an elementary school program. It can offer elementary children many experiences that are essential for attaining formal thought. Let's look briefly at some characteristics associated with each stage of development.

SENSORIMOTOR STAGE

At birth, an infant does not possess even the simplest sensory or motor skills. No directed, purposeful motor activity or well-focused sensory activity is observable in the newborn human. But the capacity to develop these basic skills is present.[3] Reflex action and random movement characterize the early part of this stage. Gradually, by interaction with the environment and the passage of time, the young child begins to develop control of motor and sensory skills. During this stage, the child cannot "separate thinking from external action; he 'thinks' in external action."[4] That is, he cannot think about actions prior to carrying them out.

As children progress through the sensorimotor stage, they develop these abilities:

1. The ability to focus on an object
2. The ability to move toward an object in a coordinated manner
3. The ability to manipulate an object
4. The ability to repeat an action

These actions reflect the development of what Piaget calls sensorimotor schemes. Such schemes are not present at birth but are the product of time and children's interactions with their environments. The above abilities are also dependent upon outside stimuli.

---

[2]Bärbel Inhelder and Jean Piaget, *The Early Growth of Logic in the Child* (New York: Harper & Row, 1964).

[3]Hans G. Furth, *Piaget for Teachers* (Englewood Cliffs, N.J.: Prentice-Hall, 1970).

[4]Furth, p. 25.

Children in the sensorimotor stage are easily distracted by new stimuli and quickly forget original intentions. When children begin to exhibit behavior that indicates greater attention and more goal-directed activity and when they can initiate some action of their own, they are beginning to move out of the sensorimotor stage.

PREOPERATIONAL STAGE

The first characteristic of children who have progressed beyond the sensorimotor stage is the recognition of object permanence or that objects exist even when they can't be seen or touched. Children who have developed this capability will look for a desired object when it is hidden from them.

Another characteristic of the preoperational stage is symbolic behavior. A child begins to demonstrate things through symbolic actions without depending on physical events. For example, she may pantomime the actions involved in eating without actually eating. It is this capacity to *think* about actions which distinquishes a preoperational child from a sensorimotor child.

Verbal language development also takes place during the preoperational stage. Children are now able to describe their thoughts and the things around them. They base their thinking on their own personal perspectives and experiences. However, they cannot see things from another's point of view; so their behavior reflects a self-centeredness. This should not be interpreted as selfishness. It means that a child's understanding or knowing is dependent upon personal experience and background. For example, a child who has a pet dog has a different understanding of the word *dog* than a child who has only seen pictures of a dog.

There are other characteristic behaviors of preoperational children. Children in this stage focus on only one property or variable so as to exclude any others, give contradictory and/or magical explanations, are dependent upon trial and error for most actions, and lack the ability to reverse actions mentally.

CONCRETE OPERATIONS STAGE

Typically, children between the ages of seven and twelve acquire the ability to perform elementary logical operations, but only through concrete means. They are unable to engage in hypothetical reasoning but can perform *mentally* what has previously been performed *physically*. Because of this newly acquired ability, the concrete operational child, unlike the preoperational child, can reverse actions mentally.

During this stage, a child can be given two identical containers of juice with exactly the same amount of liquid in each. Then the juice in one of the containers can be poured into a much shorter but wider container. Ask the child, "Is there more, less, or the same amount of juice in the two containers?" A concrete operational child who has physically experienced the reverse action should be able to mentally reverse the action and reply that there is the same amount in both. If the child has not concretely experienced the reversal, he will probably reply differently. When the child is given the opportunity to experience physically the reversal action—pouring the juice from the shorter, wider container back into the original one—he sees that the amount is the same as before and equal to the other. With many experiences of this kind, children begin to develop the concept of conservation (usually in this order: number, matter, length, area, weight, and volume).

At the concrete operational stage, this child will know if the beakers contain the same quantities of juice.

During the concrete operational stage, children also develop the ability to isolate variables and are able to think in steps without relating each step to all the others. They begin to be aware of contradictions and will try to resolve them. However, concrete operational children cannot yet go beyond that which is empirically given and are able to deal only with ideas and thoughts that result from direct personal experience. The ability to *think* about one's own thought is not yet present.

FORMAL OR ABSTRACT THOUGHT

Typically, a child begins to develop the ability to think in abstract terms—beyond one's own personal experiences—around the age of eleven or twelve. The thought processes begin to be markedly different from those of a concrete operational child. At the formal stage, the child can carry out *mental* experiences as well as actual ones.

At the formal stage, children experience mentally.

She can deal with the possible and is not satisfied with simply the empirical event given. Deductive reasoning—the ability to consider all possible combinations—and controlled experimentation are ably performed at this stage.

Summary

The ability to think and learn is itself a growing thing. Therefore, we must work at helping children develop their mental capacities. If we wish to help children achieve formal thought, we must provide many opportunities for concrete experiences. This philosophy is reflected in the new approach to science. Elementary children are involved in learning experiences in which they utilize process-inquiry skills in gathering, organizing, analyzing, and evaluating science content. This new emphasis is reflected in a change of name—from the noun *science* to the verb *sciencing*. In the following Parts, you will be involved in sciencing.

The first seven activities in this section were designed to give you some appropriate practice in using the basic process skills of observing, classifying, measuring, using spatial relationships, communicating, predicting, and inferring. Activity 8 involves you with the integrated processes of defining operationally, formulating hypotheses, interpreting data, controlling variables, and experimenting. Since each of these skills was briefly defined in the introduction to Unit 1, you may want to refer to it while completing these activities in order to satisfactorily mesh the theory with the application.

Activities 1 through 7, which deal with the basic processes, may be completed in any order you choose, but Activity 8, which involves the integrated processes, should not be done until all of the first seven are completed. Your instructor should be available for feedback. It is expected that you will maintain a continuing dialogue with the instructor as well as with your classmates. This constant interaction is an essential ingredient to the successful attainment of the specified competencies.

Space is provided for recording your reactions, questions, and comments to the various activities. (If necessary, use additional paper.) This written record will provide valuable insights and lend structure to the many small group discussions and the final seminar included in Part 1-1.

Activity 1: Observing

A. In this activity you are to select an object from those provided and list at least ten properties of that object that you can observe. Record your answers and reactions. You may work with a partner or small group in completing this activity.

The criteria presented below can be used to evaluate the quality of observations.

√ SELF-CHECK

1. Review your list to see that it contains *only* those properties that can be directly observed. In other words, you must have *firsthand knowledge* of all properties listed. For example, if you listed "it will float," you must have *directly* observed this. You must not use past experience. Cross out any properties listed that you did not directly observe.

2. Now check your list to see if you used all of your senses in making observations about your selected object. Most people tend to rely on sight observations. If you find your list heavily weighted toward sight observations, try using your other senses of hearing, touching, and smelling. You may omit tasting the object. Remember, even if the object makes no sound while stationary, that is an observation! Make additional observations and add them to your list.

3. Did you include an observation which required quantitative measurement? How long is it? How wide? How much does it weigh? Check over your list. If you find you did not include an observation based on quantitative measurement, please add one now.

4. Do your qualitative observations have a reference point? If you listed qualitative properties such as large, hard, smooth, etc., you need to give a reference point. Larger than what? Harder than what? Smoother than what?

5. Make sure that you have included at least one observation in which you acted on the object. Will it roll? Does it burn? Will it dissolve? If you did not, add that now.

It should be clear to you now that there really is a difference between making observations and making good observations. You may wish to select another object and try listing your observations again now that you are more aware of the criteria for good observations. You are also encouraged to share your experience with your classmates and exchange ideas and insights gained in this activity. Your instructor is also available to give help and feedback if you need it, or to offer encouragement and moral support.

B. The second part of this observing activity is designed to give you more practice in using your senses to gather information. Since people tend to rely on their sense of sight so much, other senses become less useful. In fact, many people find it very difficult to accurately describe an object without the benefit of seeing the object. Therefore, in the second part of this activity, you are to select a container from those provided and list as many observations as you can (at least seven) about the object(s) *inside* the container. The container is sealed, and you are not to look inside it or damage it in any way. Use your senses other than sight and the criteria from the self-check in making observations. *Remember to list only those properties that can be directly observed.* Do not list inferences or conclusions.

Container selected:

Observations:

This student is using his sense of hearing to make observations.

SELF-CHECK √

Make sure no observations are included which would require the use of your sense of sight. You may not, for example, include an observation such as, "the object in the container is round." This observation would require you to actually see or feel the object. Neither of these avenues are available to you since the object is sealed in an opaque container. What you can observe are sounds made by the object(s), the weight of the object(s) plus the container, and other observations made by utilizing your senses of hearing, touching (through the barrier of the container), and smelling. Cross out any observations that do not meet the criteria and add any necessary additional observations. Compare your observations with others and jot down any problems, questions, or comments that you want the instructor to explain or discuss during the small group seminar to be held when you complete all eight of these activities. When you feel that you have achieved competence in observing, go on to one of the other activities.

COMMENTS

Activity 2: Classifying

A. In this activity you will select a container from those provided and separate the contents into groups. You are to identify and label each group according to the property you used to separate the groups.

1. Separate the contents of the container you selected into two groups. Record the property used to make the separation below.

2. Can you find another way to separate the contents of the container into two groups? Label the two new groups.

3. Take about three minutes and see how many different ways you can divide the contents of the box into two groups. Label each group.

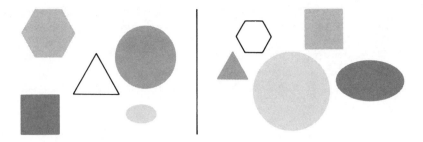

SELF-CHECK √

1. Check over your list to see that all the objects in the container fit into one of the groups in each set. But make sure that no *one* object could fit into both groups of one set. You might want to have a classmate check over your list while you check his or hers. Discuss any problems you are having and ask your instructor for feedback.

2. Look over your sets of groups to see if you have any set which would allow for the inclusion of any additional objects not included in the set of objects provided for you. Put an asterisk beside those groups. If none of the sets of groups formed thus far would allow for any other objects to be included, then form two groups which *will* allow for the inclusion of any additional object. Use the space below to record your response.

B. Now that you have had some practice in grouping, or classifying, objects into two sets, try to separate the contents of the container into three groups. Make as many sets of three as you can and label them in the space below.

C. Compare your responses with those of your classmates. Use the same criteria given for the groups of two to evaluate the quality of your work.

1. If you were trying to help children develop the skill of classifying, what questions might you ask to encourage the learners to investigate the various properties that could be used in grouping the objects?

2. Jot down questions and comments for later discussion.

COMMENTS

Activity 3: Measuring

Linear measurements can be made without a ruler.

A. This activity will involve you in the process of measuring. You will make linear measurements, liquid measurements, and mass/weight measurements. Work with a small group in completing this activity.

Linear Measurement

1. Select a nonstandard measuring device, identify it by some means, and have each group member measure the same object or distance in the classroom using the selected device. You may choose a device from those provided by your instructor or create your own. Body parts such as hands, feet, or arms make good measuring devices. Record your data below.

a. Identify your measuring device.
b. Identify what was measured.
c. Identify the distance measured.
d. How does your measurement compare with those of the other members of your group?

1. Is the recorded measurement consistent with the type of measuring device used? For example, if you selected a paper clip as the measuring device, then the object or distance measured should be recorded as so many paper clips in length.
2. Did everyone in the group use the same device for measuring? For example, if your group chose to use a hand as the measuring device, then the same hand must be used by all. Hands come in different sizes, as do other devices that might be used in measuring distances. It is very important that a standard be established so that measurement can be related.

Liquid Measurement

1. Select two containers from those provided. One should be much larger than the other. Identify each by some means and have each group member use the smaller container to determine the amount of liquid that the larger container holds. Record your data below.
a. Identify your measuring device.
b. Identify what was measured.
c. Identify the amount measured.
d. How does your measurement compare with those of the other members of your group?

SELF-CHECK √

1. Is the recorded measurement consistent with the type of measuring device used? For example, if you identified the measuring device as "a small orange juice can," then the amount recorded should be in so many small orange juice cans.
2. Did you establish a way to handle the problem of describing the amount measured when the measuring device was only partially filled? If not, do so now.

Mass/Weight Measurements

1. Select an object and measure its mass/weight using the simple equal arm balance provided.
2. Place the object to be measured on one side of the balance and measure its mass/weight by placing paper clips on the other side until it balances. Record your data below.
a. Identify the object to be measured.
b. What was the mass/weight of the object?

Using paper clips as a standard, these children compare the mass/weight of various objects.

√ SELF-CHECK

1. Is the mass/weight of the object recorded so that it is clear paper clips were the standard?
2. What other things could you use as a standard for measuring mass/ weight?

Discuss the following with your small group when you have completed the three parts of this activity:

1. How were you involved in the process of measuring?
2. Can you explain this process?
3. How important is it to identify a standard when making measurements?

You may use the space below to record comments, questions, and conclusions that resulted from your small group discussion. You may also want to invite your instructor to join your discussion group if you feel you need more feedback on this activity.

COMMENTS

Activity 4: Using Spatial Relationships

A. In this activity, you will be working with four geometric shapes—squares, rectangles, parallelograms, and triangles of different sizes.

1. Using the set of seven shapes provided, select any two and produce a third geometric shape, that is, either a square, rectangle, parallelogram, or triangle. Draw the shape indicating component shapes with dashed lines.

2. You might want to try some other combinations of two shapes to see the result. Draw your results.

3. This time, use four of the shapes to construct a square, rectangle, parallelogram, or triangle. Draw your results below.

4. Try some other combinations of four shapes. Draw your results.

5. This one will be a little bit harder, but it *can* be done. In fact, there are several ways to do it. Take all seven shapes and construct either a square, rectangle, parallelogram, or triangle. Don't give up. Try to apply what you learned in the first two parts of this activity. If you have trouble, try making half a rectangle or square, then match the remaining shapes on top of these. The bottom part will give you a concrete picture of what you need to form the entire rectangle or square. Then just slide the top pieces off and you should have the whole shape. Work on this first by yourself, then with another person or small group if you have trouble. Draw your completed shape in the space below and indicate the component shapes with dashed lines.

This child is learning that geometric shapes can be constructed from other geometric shapes.

B. Discuss any difficulties and frustrations with your classmates. Share your hints with a classmate who is having problems with this activity, but don't show how you did it! The space below is for recording comments and questions for later discussion.

COMMENTS

Activity 5: Communicating

This activity is designed to extend the skills acquired in the observing activity. If you have not completed Activity 1, it is suggested that you do so before beginning this activity. *Communicating* in science refers to the skill of describing simple phenomena. A written or oral description of physical objects and systems and the changes in them is one of the most common ways of communicating in science. Constructing graphs and diagrams for observed results of experiments is another form of communicating.

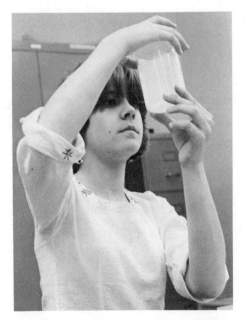

Learning how to describe what you see is important in science.

A. Using the substance provided, describe it by listing at least ten observations. *Remember the criteria for good observations.* Record your responses below.

√ SELF-CHECK

1. Does your list include observations in which you used four of the five senses? (exclude taste)
2. Did you act upon the object?
3. Did you include an observation in which you gave some quantitative measurement?

B. Place the substance in one of the containers filled with liquid that are provided and observe. Describe as many physical changes as you observe.

√ SELF-CHECK

1. Did your observations include all of the criteria mentioned in the first part of this activity?
2. Did you include changes in the liquid as well as in the substance?

C. Describe any characteristic that you observed which remained unchanged.

√ SELF-CHECK

1. Did you include the state of the liquid, its color, level, temperature?
2. Compare your reactions with those of others and discuss the application of this skill for elementary children. Record any comments or questions below.

COMMENTS

Activity 6: Predicting

Making a prediction is very different from just guessing. Predictions should be based upon selected data. Two types of predictions are possible using graphically presented data: (1) interpolation, within the data, and (2) extrapolation, beyond the data. In both types of predictions, data are gathered and recorded in graph form. A pattern should emerge and the prediction is made. The following activity is designed to help you engage in the process of predicting.

A. Use the set of beakers provided. You will notice that they are labelled 1, 2, 3, and 4. A large candle, matches, and several smaller candles are also provided. Light the larger candle and use it in lighting the smaller candles throughout the activity. This way you will conserve the number of matches you use.

These students are using a process-inquiry activity to learn to make predictions.

1. Using the lighted larger candle, light one of the smaller candles and place Beaker 1 over it. Record how long the candle will burn. Make three trials and obtain an average time. Record the data below.

Measurements for Beaker 1

first trial   second trial   third trial   average

2. Relight the smaller candle and place Beaker 3 over it. Be sure you use Beaker 3. Measure the burning time and record below as for Beaker 1.

Measurements for Beaker 3

first trial   second trial   third trial   average

3. Use the bar graph below to record your data for Beaker 1 and Beaker 3. On the left side, the time is indicated so that each set of horizontal blocks represents a specific number of seconds. Space is left at the top for you to continue the graph if necessary. The grid provides space for recording the burning time of the candle for each beaker in bar graph form. Notice that space is provided for both *predicted* and *actual* time for Beakers 2 and 4. You will be able to see the results better if you choose a different color for marking the predicted and actual times.

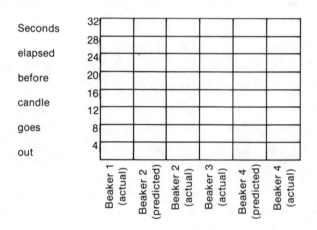

4. Using the data gathered thus far, *predict* the length of time the candle will burn under Beaker 2. Record this prediction below and on the graph. Tell how the data were used to make your prediction.

   a. Predicted time the candle will burn under Beaker 2:

   b. How was the data used to make your prediction?

5. Light the smaller candle again and record the actual time it takes for it to go out when Beaker 2 is placed over it. Make three trials and then average the results. Record the data below and on the graph.

Measurements for Beaker 2

first trial   second trial   third trial   average

6. Using the data gathered, predict the length of time the candle will burn under Beaker 4. Record your prediction below and on the graph.

   a. Predicted time the candle will burn under Beaker 4:

   b. How was the data used to make your prediction?

7. Light the smaller candle and measure the actual time it takes for it to go out when Beaker 4 is placed over it. Record the data below and on the graph.

Measurements for Beaker 4

first trial   second trial   third trial   average

SELF-CHECK √

Review your results. Are they recorded so that others could understand them? Check your predictions against the actual measured times. How close were they? If there is more than two seconds' difference, how do you account for the discrepancy? What could you do to make your predictions more accurate? Record your response below.

COMMENTS

Activity 7: Inferring

Inferences differ from observations. An inference is a conclusion or judgment based upon observations. It is arrived at *indirectly* rather than directly. For example, you see the sun shining on your car. You know that your car has been parked in the sun for three hours. You can feel that the air around you is very warm. You *infer* that the hood of your car will feel very warm to the touch. You have not experienced this directly as yet, but based upon your observations, you can infer that the hood will feel warm when you touch it. In this example, you can test this inference by actually touching the car.

In this activity you will use the same sealed containers that are used in the observing activity. This time you will make *inferences* about the object(s) inside the container. In order to help you become more aware of the process of observing, you will also list the properties *observed* that you used in making the inferences.

*These students are learning
to make inferences from
what they observe.*

1. Select two of the sealed containers used in the second part of Activity 1 and make inferences about the object(s) concealed in the container. *Remember, inferences should be based upon the observations you make.* Record your inferences and observations below.

   a. Container_____

   Observations   *Inferences*

   b. Container_____

   Observations   *Inferences*

√ SELF-CHECK

1. Were your inferences based upon the observations you listed? If not, how did you determine the identity of the concealed objects?

2. If someone else read your list of observations, could he make the same inference(s) that you made about the object's identity?

3. Compare your results with others and discuss any problems or questions that arise. Record your comments or questions for later discussion.

COMMENTS

You have now completed all the activities involving the basic processes. These activities were all designed to give you some first-hand experiences with the basic process-inquiry skills. For the most part, the activities were void of content in order to help you see clearly the process involved. Science activities that you plan for children will probably include content as well as process. The two should be meshed in order that elementary children not only understand the content of science, but also develop facility in the use of the process skills in investigating natural phenomena. You should examine science curricula materials carefully to see that process is included and that science is not merely memorizing facts and repeating them back on a test. Piaget tells us that children need first-hand experiences in which they are involved in using the process skills in order to advance from one developmental stage to the next and to reach the final stage of formal thought.

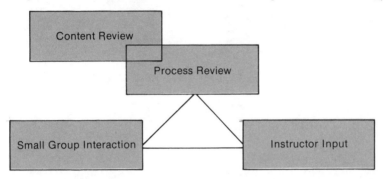

Review the activities so far. Form a small group with your classmates and discuss the completed activities. Share your reactions with your instructor. If you feel that you have mastered the basic process-inquiry skills, you may do Activity 8.

Activity 8: Integrated Processes

In this activity, you will utilize the necessary processes involved in designing and conducting an experiment to answer a selected question. The extent to which one is able to utilize the integrated processes (defining operationally, formulating hypotheses, interpreting data, controlling variables, experimenting) depends upon one's stage of mental development. Most elementary children will not be able to utilize the integrated processes to a great extent. This does not mean that elementary children should not be provided with concrete experiences in which they utilize the various integrated processes. It does mean that the teacher must be selective and structure the experience so that elementary children work with limited variables and are given the opportunity to manipulate concrete materials in solving problems.

With one or two classmates, select a question from those provided and design and conduct an experiment to answer it. Decide on operational definitions, if any are needed, to clarify terms that will be used in answering the question. Use the space provided to record your plans for solving the problem.

Applying their new skills, these students conduct an experiment.

1. Method or strategy to be used in answering question.

2. Now you need to construct a hypothesis (a tentative answer to your selected question based upon previous knowledge and experiences) to be tested. State your hypothesis below.

3. In designing the experiment, you must consider *all* variables that could affect the outcome or results. List all possible variables.

4. In order to conduct the experiment, you must manipulate only one variable at a time. The other variables must be controlled or the outcome may be contaminated. Identify the one variable that you feel is most likely to cause the outcome you indicate in the hypothesis.

5. You must now design your experiment so that all the other variables are controlled. Record the data gathered in the experiment.

6. Now review your data. Relate this data to your stated hypothesis. Is the data you gathered sufficient to answer the selected question? Is your hypothesis supported or not supported? If your data are not sufficient to answer your question, you must continue the experiment using the other possible variables. Remember to allow only one possible variable to vary at a time. You might also examine your hypothesis and re-evaluate your design. If you have trouble with this, ask the instructor for help. If your data are sufficient to answer your question, then you can make generalizations and conclusions based upon the gathered data. State your generalizations and conclusions below.

7. Indicate what implications you feel there might be for further experimentation.

8. Arrange for your instructor to meet with your group for a brief presentation of your experiment. This presentation should include a description of the method or strategy used, hypothesis constructed, variables considered, variable manipulated, data gathered, and generalizations and conclusions reached. A demonstration involving the equipment and materials used in conducting the experiment should also be included.

COMMENTS

Summary: Sciencing

In completing this Part, you were involved with ideas, materials, and activities designed to help you acquire specific information and the skills necessary for achieving competence in sciencing. The background information provided you with insight into the nature of learning and helped establish a rationale for the new approach of process-oriented science. The activities were designed to help you become more aware of the specific process skills. In doing them, you used these skills in gathering, organizing, analyzing, and evaluating data.

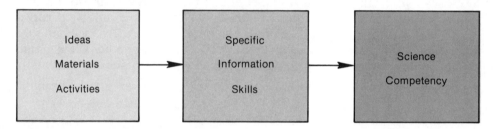

Science is no longer defined as a body of knowledge but takes its definition from what scientists do. The process-inquiry skills are the tools of scientific investigation, which are used to gather, organize, analyze, and evaluate science content. When you have completed this Part, you should have become much more aware of the importance of process skills and the role they play in all later learning.

FINAL SEMINAR

Review this Part and look over the questions and comments you recorded after the various activities. Get together with a small group of your classmates and discuss your reactions to this Part. If you have specific questions, share them with your group. Together you might be able to answer them. *Make sure your instructor is invited to share in this final group seminar.*

COMPETENCY EVALUATION MEASURE

The activites you have completed are considered appropriate practice. In order to evaluate your competency in the area of sciencing, your instructor may choose to use a *competency evaluation measure* of some type. Consult with your instructor for specific directions.

BIBLIOGRAPHY

Bloom, Benjamin S.; Hastings, Thomas; Madaus, George F. *Handbook on Formative and Summative Evaluation of Student Learning.* New York: McGraw-Hill Book Co., 1971.

Carin, Arthur, and Sund, Robert. *Teaching Science through Discovery.* 3d ed. Columbus, Ohio: Charles E. Merrill Publishing Co., 1975.

Esler, William K. "Putting It all Together—Inquiry, Process, Science Concepts, and the Textbook." *Science Education,* 57, no. 1 (1973); 19-23.

Furth, Hans G. *Piaget for Teachers.* Englewood Cliffs, N.J.: Prentice-Hall, 1970.

Gega, Peter C. *Science in Elementary Education.* New York: John Wiley & Sons, 1970.

George, Kenneth D. *Elementary School Science: Why and How.* Lexington, Mass.; D. C. Heath & Co., 1974.

Inhelder, Bärbel, and Piaget, Jean. *The Early Growth of Logic in the Child.* New York: Harper & Row, Publishers, 1964.

McAnarney, Harry. "What Is the Place of Product and Process in the Development of Generalizations in Elementary Schools Science? *"Science Education* 56, no. 1 (1972): 85-88.

Piltz, Albert, and Sund, Robert. *Creative Teaching of Science in the Elementary School.* 2nd ed. Boston: Allyn and Bacon, 1974.

SUGGESTED READING

Bloom, Benjamin S. et al. *Taxonomy of Educational Objectives, Handbook I: Cognitive Domain.* New York: David McKay Co., 1956.

Carin, Arthur, and Sund, Robert. *Teaching Science through Discovery.* 3d ed. Columbus, Ohio: Charles E. Merrill Publishing Co., 1975. Chapters 1, 2 and 5.

Esler, William K. "Putting It all Together—Inquiry, Process, Science Concepts, and the Textbook." *Science Education* 57, no. 1 (1973): 19-23.

Furth, Hans G. *Piaget for Teachers,* Englewood Cliffs, N.J.: Prentice-Hall, 1970.

Gega, Peter C. *Science in Elementary Education.* New York: John Wiley & Sons, 1970. Chapter 1.

George, Kenneth D. *Elementary School Science: Why and How.* Lexington, Mass. D. C. Heath & Company, 1974. Chapters 1 and 2.

Harrow, Anita J. *A Taxonomy of the Psychomotor Domain.* New York: David McKay Co., 1972.

Krathwohl, David R.; Bloom, Benjamin S.; and Masia, Bertram B. *Taxonomy of Educational Objectives, Handbook II: Affective Domain.* New York: David McKay Co., 1964.

UNIT 2
Science Curricula Material

In recent years, much effort has been directed toward making science more relevant to the modern world and toward helping teachers do a better job of teaching science. Millions of dollars have been invested in elementary science curriculum projects by governmental agencies and private corporations. Three types of teaching programs have evolved.

Three Types of Programs

TEXTBOOK APPROACH

The textbook is a "first generation" or traditional content approach to teaching science. A textbook serves as a curriculum guide, a reader, and a resource. Current texts also stress hands-on science activities and science processes. Some examples of this approach are those texts published by Scott, Foresman and Co., Ginn and Co., and Charles E. Merrill Publishing Co.

LABORATORY APPROACH

The laboratory approach (or second generation approach) is a kit that usually contains all of the materials needed to teach the program, including an instructor's manual. Very few reading materials are provided for the student. Science processes or "how to do science" is stressed. The child learns by doing, not by reading about science. Examples of this approach are SCIS and *SCIIS* by Rand McNally & Co., SCIS II by American Science and Engineering (AS&E), ESS by McGraw-Hill Book Co., and SAPA II by Xerox, all of which will be discussed in this text.

TEXT-KIT APPROACH

The text-kit approach (or third generation approach) has grown out of the previous two approaches. The basic security and guidance provided by the textbook are combined with the hands-on activities of the science kits. Usually, kit materials are simple, readily available materials that can be obtained locally or purchased from the publisher. Examples of this approach are M.A.P.S. by Houghton Mifflin Co. and STEM by Addison-Wesley Publishing Co., also to be discussed later.

The three Parts which follow will introduce you to these approaches. They do not have to be done in any particular order. Sample activities from each type of program will enable you to obtain a hands-on understanding of the materials used and the methodology employed. In completing these Parts, you will be able to do the following:

1. Identify and describe the major science programs available at the present time
2. Analyze and evaluate the various science programs available at the present time
3. Utilize the various teaching strategies found in each of the major science programs

PART 2-1

Laboratory Approach to Elementary Science

FLOWCHART: Laboratory Approach to Elementary Science

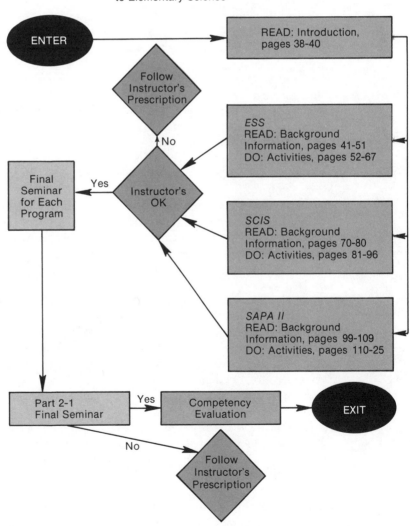

Introduction

This is the first of three Parts that will introduce you to science curricula materials. In this Part, you will be actively involved in examining, analyzing, and evaluating three of the better-known commercial kits:

1. *Elementary Science Study (ESS)*
2. *Science Curriculum Improvement Study* (SCIS, *SCIIS*, and SCIS II)
3. *Science . . . A Process Approach II* (SAPA II)

You may wonder why it is necessary to study these specific kits since you may not have any of them where you go to teach. True, you may not have access to any of them, but that is not important. What is important is that you will learn some unique teaching strategies. You will also be aware of some of the science programs that are available to the public schools. Furthermore, you will be exposed to a variety of activities that you might be able to adapt to your classroom. Remember, these programs were developed as a result of a desire by science educators to break away from the traditional methods of teaching elementary science. They are based on recently developed concepts in child development. Therefore, they contain unique ideas and teaching strategies. These ideas and teaching strategies are what you are going to learn.

This Part is divided into sections covering the three programs mentioned above. As you review the three laboratory approaches to science, you will find that they are quite different from each other. At the same time, you will find that there are common bonds that produce similarities. The differences will become apparent as you work your way through the text, but the similarities are less apparent. Let us look at some of these similarities.

All three of the programs share a common origin. After Russia launched its Sputnik in 1957, the immediate reaction in the United States was to question our science programs, from kindergarten through college. As a result, we found out that most science programs, especially elementary programs, were inadequate, that most teachers did not have much science background, and that textbooks were generally obsolete.

The National Science Foundation (NSF) was established by the federal government to provide financial help for science projects and to improve science education. Many ideas and projects designed to improve science in the elementary school were advanced by many sources. Individuals, public school systems, universities, private companies, and scientific organizations were funded to develop programs. Some were successful, but a great number were not. Three programs—*Elementary Science Study* by the Education Development Center, *Science Curriculum Improvement Study* by Robert Karplus, and *Science . . . A Process Approach* by the American Association for the Advancement of Science—emerged as the most successful. It is interesting to note that of these programs, one, ESS, was developed by a private company; another, SCIS, by a private individual working with a local school system; and the third, SAPA, by a scientific organization.

The National Science Foundation recognized the fact that even the best science program is of little value if it is not available to the public. Consequently, private publishing companies were invited to bid for the rights to market many of the science programs. The company would publish, promote, and market the program and, in return, would be given exclusive marketing rights for a predetermined period of time, then the programs would be in the public domain. Of all of the programs funded by NSF, only three, ESS, SCIS, and SAPA, were accepted for commercial publication.

Upon examining these three programs, several common characteristics become apparent. First of all, three distinct approaches to teaching elementary science emerge. They are very different, but one goal is common to all three: Children should learn how to *do* science, not memorize subject-matter content. The science processes are stressed as a way to accomplish this goal. The underlying assumption for these programs is that every person, child or adult, uses science daily whether they know it or not and should be prepared to meet any problem they encounter.

A second common characteristic is the use of a hands-on, laboratory approach teaching and learning science. Work by Piaget and other psychologists indicated that children in the elementary school learn best by manipulating concrete objects. So, all three programs try to get materials into the hands of children as quickly as possible for concrete experiences, and reinforcement. Involvement with the materials allows children to be partners in the learning process by actually doing what scientists do.

Many children have reading problems which penalize them in all of their school work. They can't read the materials; therefore, they can't do the work. This problem is minimized in a hands-on laboratory approach. Children who cannot read can participate in almost all of the activities. They learn by doing, rather than by reading. Success in science might encourage some of the slower readers to want to learn to become better readers so that they might supplement what they are learning.

Each project developed a unique teaching strategy but shared one characteristic. The roles of the teacher and student changed. The teacher became a guide and resource rather than a dispenser of knowledge. The student became an active participant rather than a passive receptor in the learning process. This creates one major problem: in general, teachers don't know how to teach this way. They must be retrained to teach these programs properly. More than one of these programs have failed because the teacher did not know how to use it. Preservice and inservice workshops, which may vary in format or time needed, are recommended for all three programs. When the teacher knows his or her role, the students can perform accordingly.

You are now ready to review the ESS, SCIS, and SAPA II programs. As you go through them, try to see how they are alike and how they are different. Mentally compare them and think about what each has to offer the classroom teacher and, maybe more important, what each has to offer the students. In working through Part 2-1, you will follow a sequence of five steps:

1. Read the background information. This will help you become familiar with the history, philosophy, and design of the program under consideration. Remember, read *everything* before starting anything.
2. Review the teacher's manuals or guides from each program. These may be obtained from your instructor. This is the first step in getting your hands on the actual material used by the particular program. The guides will give you insights not available from background readings. Don't try to see how many you can look at or how fast you can look at them. A small number of manuals seriously reviewed will be much more beneficial.
3. You will actually do some of the activities found in the program under consideration. These activities have been selected to give you a sample of the types of activities used and how they are used. They have been modified somewhat to make them appropriate for your use and to make them usable in a shorter time

span, but they still reflect the teaching strategy of the program from which they came. Your hands-on involvement in the actual teaching materials should add to your insight into the program.

4. You will be involved in self-checking and instructor feedback situations throughout. At various stages, you will be presented with self-check questions and suggestions to help you judge your progress. The instructor will also be available to answer your questions or to direct you to new lines of thought as you look into the programs.

5. You will participate in a competency appraisal based on your understanding of the program. The instructor will determine the format of this evaluation.

As you go through these steps, you need to look for some specific information:

1. What is the basic teaching strategy, and how can you adapt this strategy to your teaching repertoire?

2. What kinds of concrete materials are used in the activities? Could you obtain the same or similar materials for your classroom?

3. What do you think of each program as a whole? Would you like to teach that program? Why or why not?

4. How could you adapt some of the activities to your classroom? In what situations could you best use each of them? How would you teach them?

5. In Unit 1, learning the process skills and sciencing were heavily emphasized. Can you find evidence of these concepts in the three programs you are reviewing? What implications does this have for your personal teaching strategy?

If you will seriously look for answers to these questions while you are doing this Part, you will learn much more than just the names of some science programs. You will learn new ways to teach science to elementary children. As you go through this Part, you will be involved in some individual, some group, and some individual-group instructor activities. Each type of activity should help you learn and reinforce what you learn. Don't try to do the entire Part alone. Feedback and interaction are very important. Remember to read the background material before you start an activity and to read the directions before you try to answer the questions.

GOAL

After completing this Part, you will demonstrate competency in the identification and utilization of equipment and curriculum materials that could be used in a laboratory approach to elementary science.

BEHAVIORAL OBJECTIVES

In completing this Part, you will do the following:

A. Identify and describe the materials in the Elementary Science Study (ESS), *Science Curriculum Improvement Study* (SCIS) and its revisions *SCIIS* and SCIS II, and *Science . . . A Process Approach II* (SAPA II) programs.

B. Identify and describe the scope, sequence, and teaching strategy of ESS, SCIS and its revisions *SCIIS* and SCIS II, and SAPA II.

Elementary Science Study (ESS)

Background Information

Author: Developed by Education Development Center
           55 Chapel Street
           Newton, Massachusetts 02160

Publisher: Webster Division
           McGraw-Hill Book Company
           Manchester Road
           Manchester, Missouri 63011

History

In 1958, the Education Development Center was formed as a nonprofit organization devoted to the improvement of instruction in schools. It was originally funded by grants from the Alfred P. Sloan Foundation and the Victoria Foundation, and later by continuing grants from the National Science Foundation. In 1960, small-scale work began on a science program to be used from kindergarten through the eighth grade. This grew into the present *Elementary Science Study* (ESS) materials. These materials are constantly revised, although no new units have been developed recently. New supplementary materials, based on research projects and classroom usage, are either incorporated into existing materials or published as supplemental materials.

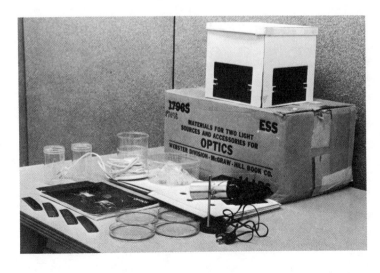

Basic equipment for the ESS Optics activity.

The developers (there are no authors as such) of the ESS program came from all sectors of the scientific and teaching communities. Chemists, physicists, mathematicians, biologists, engineers, and teachers representing every level of public and private education from kindergarten to the university contributed to the program. Probably the most important developers were the school children who tried and criticized the ideas and materials presented to them.

There were several steps in the procedure for developing a unit. Someone, either a teacher or scientist, would get an idea for a unit that they would then try in a classroom. Critiques by the users—teachers and students— helped to either discard the idea or develop it further. After numerous critiques and rewritings, the final unit emerged for publication. Eventually, fifty-six units, which included physical science, biological science, earth science, and mathematics, were developed. Constant feedback by users has helped keep the program current.

Conceptual Background

The conceptual framework for ESS is based largely on the work of Jerome Bruner.[1] He stressed that every topic can be taught in some intellectually honest form to any student. The secret is to match the level of the student and the level of the material. He also stressed the act of discovery, which he defined as obtaining knowledge for oneself by the use of one's own mind. The teacher is considered, by Bruner, to be the key to learning. It is the teacher who provides the opportunity for learning to occur. All of this suggests that a good science program should have a teacher who can use almost any topic to provide the students with the opportunity to discover.

Program developers, from the very beginning, insisted upon "units which satisfied two criteria: their scientific content is significant, and the activities, materials, and subject matter make children curious about some part of their world and encourage them to learn more about it."[2] Simply put, elementary science should be fun but should also be informative and accurate.

The following five major goals found in the ESS program reflect a blending of cognitive, affective, and psychomotor skills:

I. Rational Thinking Processes
   A. Observation
      The student will develop skills in the identification of objects and their properties, changes in properties, controlled observations, and ordering of a series of observations.
   B. Classification
      The student will develop skills in classifying objects, actions, and phenomena.
   C. Measurement
      The student will demonstrate the ability to measure length, area, volume, weight, temperature, force, and speed.

---

[1]Jerome Bruner, "The Act of Discovery," *Harvard Educational Review 31* (Winter 1961): 21-32.

[2]Elementary Science Study, *A Working Guide to the Elementary Science Study* (Newton, Mass.: Education Development Center, 1971), p. 2. Available from the EDC Distribution Center, 39 Chapel St., Newton, Mass. 02160.

D. Data Collection and Organization

The student will organize information pertaining to a scientific investigation, describe it verbally or graphically, and present the data in such a way that trends can be analyzed.

E. Inference and Prediction

The student will predict an outcome from a trend in data using inference, extrapolation, or interpolation.

F. Variable Identification and Control

The student will identify the independent and dependent variables in experiments and describe how variables are controlled and interrelated.

G. Making and Testing Hypotheses

The student will identify a scientific question, describe procedures to be used toward answering it, carry out the procedure, and evaluate the results.

H. Process Synthesis

The student will utilize all rational thinking processes by designing and carrying out a scientific investigation and reporting its results.

II. Manipulation

The student will be able to assemble and use the appropriate tools and apparatus needed to investigate a scientific problem.

III. Communication

A. The student will develop communication skills by describing orally, in writing, or nonverbally, his conclusions about and reaction to science and its processes.

B. The student will communicate scientific information by organizing and presenting data in graphic or mathematical symbols.

IV. Concepts

The student will be able to recall and/or apply the knowledge of facts, theories, laws, structures, or concepts of science.

V. Attitude

A. The student will demonstrate individual curiosity and persistence in the study of science.

B. The student will participate willingly in science activities, accept evidence gathered through the scientific methods, and value critical thinking.

C. The student will develop self-confidence in the study of science by being involved in a variety of activities. Through personal participation in the ESS program activities, the student will develop a positive self-image and an "I can do it" attitude toward the study of science.[3]

The conceptual framework can be summed up in a simple statement taken from the ESS *Working Guide*:

Basic threads of scientific investigation—inquiry, evidence, observation, measurement, classification, deduction—are part of the fabric of all ESS units, but they are not the whole cloth. No unit aims solely to teach individual skills, nor are any units intended primarily to illustrate particular concepts or processes or the like. Instead, by presenting

---

[3]William Aho, et al., *McGraw-Hill Evaluation Program for ESS* (St. Louis, Webster/McGraw-Hill, 1974), p. iv.

interesting problems and real materials to explore, the units invite children to extend their knowledge, insight, and enjoyment of some part of the world around them.[4]

Program Description

The ESS is different from most science programs in that it is composed of individual units. It is based on the assumption that all schools in Texas, for example, do not have the same needs as rural schools in Michigan, and neither have the same needs as large urban schools. Therefore, why assume, as most textbooks do, that one program can fit all schools? Using ESS units, each school system can tailor a program to fit its own needs. There are fifty-six units (or modules if you prefer) available, and each one can stand alone. Refer to the chart on pages 45-47 for a list of these units and the wide array of subjects to choose from. Each is designed for a grade range, rather than a specific grade level, making it adaptable to various grade levels or ability levels within a specific grade. This format allows flexibility in devising a program to fit individual needs.

The preferred implementation format is that of a total, tailored program for the school. An ESS consultant works with school personnel to help them decide which units would be appropriate to their curriculum. Usually five or six units per grade level are selected. Each teacher then teaches those units selected for her grade level. Sometimes a unit is used at a lower grade level, then repeated in more depth at another level, but, generally, units used in one grade level are not used in others. Since an orderly, preplanned program is the goal of this format, all fifty-six units are not on hand at the school and available for random selection by individual teachers.

An alternative to the complete program implementation of ESS materials is widely used. Individual teachers use various units to supplement existing programs. A teacher might prefer a complete program, but for various reasons, the change may not be feasible. A newer program might already be in existence, the school might be locked into a textbook program with little chance of change, the staff of the school may not want to change, or, as is so often the case, money may be limited. If so, teachers can obtain one or two kits and use them to supplement their materials. They can also modify their teaching strategies to that of the ESS program. Consequently, ESS can be used as a complete program or as a supplemental program and it is effective either way.

Since ESS is a kit program, usually, each unit is composed of a teacher's guide and the materials needed to teach the unit. However, this is an oversimplification because each unit is distinct. All units do have a teacher's guide, but in some units, such as *Clay Boats* or *Tangrams,* the guides are very short and simple; whereas, in other units, such as *Small Things* and *Batteries and Bulbs,* the guides are very detailed and extensive. Some units, such as *Microgardening,* have supplementary manuals for reference, and others, such as *Bones,* have picture books for discussion. Materials included in each kit also vary. A general rule is that easily obtainable items are not included, but specialized or hard-to-obtain items are usually supplied.

A unit such as *Pond Water* contains only two items: a teacher's guide and a set of pond water cards (enlarged photos of pond water microorganisms). On the other hand, a unit such as *Batteries and Bulbs* contains a teacher's guide, bulbs, wire, batteries, wire strippers, bulb holders, compasses, clips, brads, rubber bands, trays,

---

[4]Elementary Science Study, *A Working Guide . . . p. 2.*

ESS Units Grouped by Principal Subject Matter

UNITS	K	1	2	3	4	5	6	7	8	9
Biological Sciences										
Animals in the Classroom	▓	▓	▓	▓	▓					
The Life of Beans and Peas	▓	▓	▓	▓	▓					
Butterflies	▓	▓	▓	▓	▓	▓				
Eggs & Tadpoles	▓	▓	▓	▓	▓	▓	▓			
Growing Seeds	▨	▓	▓	▨						
Brine Shrimp		▓	▓	▓	▓					
Changes		▨	▓	▓	▓					
Pond Water		▨	▨	▓	▓	▓	▓	▓		
Mosquitoes				▓	▓	▓	▓	▓	▓	▓
Animal Activity					▓	▓	▓			
Bones					▓	▓	▓			
Budding Twigs					▓	▓	▓			
Crayfish					▓	▓	▓			
Earthworms					▓	▓	▓			
Small Things					▓	▓	▓			
Tracks					▓	▓	▓			
Microgardening					▓	▓	▓	▓	▨	▨
Starting from Seeds				▨	▨	▓	▓	▨		
Behavior of Mealworms					▨	▨	▓	▨		

45

UNITS	K	1	2	3	4	5	6	7	8	9
Physical Sciences										
Light and Shadows	■	■	■	■						
Printing	■	■	■	■						
Mobiles	■	■	■	■	■					
Musical Instrument Recipe Book	■	■	■	■	■	■	■	■	■	■
Spinning Tables		■	■							
Primary Balancing	▨	▨	■	■	▨					
Sand			■	■						
Structures			■	■	■	■	■			
Sink or Float			▨	▨	■	▨	▨	▨		
Clay Boats			▨	■	■	▨	▨			
Drops, Streams, and Containers				■	■					
Mystery Powders				■	■					
Ice Cubes				■	■	■				
Colored Solutions				■	■	■	■			
Whistles and Strings				▨	■	■	▨			
Batteries and Bulbs			▨	▨	■	■	■			
Optics					■	■	■			
Pendulums					■	■	■	▨	▨	▨

UNITS	K	1	2	3	4	5	6	7	8	9
Senior Balancing					▨	■	■	▨	▨	
Water Flow					▨	■	■	▨	▨	
Heating and Cooling						■	■	■		
Balloons and Gases						■	■	■	■	
Batteries and Bulbs II						■	■	■	■	■
Gases and "Airs"						■	■	■	■	
Kitchen Physics							■		▨	▨
Earth Sciences										
Rocks and Charts				■	■	■	■	■	■	
Where Is the Moon?				■	■	■	■	■		
Stream Tables					■	■	■	■	■	■
Mapping						■	■	■		
Daytime Astronomy						■	■	■		
Mathematics										
Match and Measure	■	■	■	■						
Geo Blocks	■	■	■	■	■	■	■	▨	▨	▨
Pattern Blocks	■	■	■	■	■	■	■	▨	▨	▨
Attribute Games and Problems	■	■	■	■	■	■	■	■	■	▨
Tangrams	■	■	■	■	■	■	■	■	■	▨
Mirror Cards		■	■	■	■	■	■	▨	▨	▨
Peas and Particles				■	■	■	■	■		

*The categories under which the units are listed do not encompass the total contents of the units. Activities in all of units extend into and combine elements from a number of subject areas, both in science and other studies. A few units are essentially interdisciplinary. For information about each unit, see the unit descriptions.

From Elementary Science Study, *A Working Guide to the Elementary Science Study* (Newton, Mass.: Education Development Center, 1971), pp. 8-9. Used with permission.

student prediction sheets, test cards, and project sheets. Replacement packages for the expendable items are also available. For the *Pond Water* unit, the teacher must supply everything; and for the other unit, nothing. All of the units fit between these two extremes, but the best way to really see what is in each kit is to go through an ESS catalog. They may be obtained directly from McGraw-Hill or from your local Webster/McGraw-Hill representative. Check with your instructor to see if one is available for your use.

Two ESS kits.

The cost of this program is comparable to other science programs. Some units are expensive and others are inexpensive, but a really balanced program will have some of each. The catalog gives an accurate cost analysis of this program.

Teaching Strategy

Probably the most important difference between traditional science programs and laboratory programs is in the teaching strategy employed. The teacher's role in the ESS program is very different from the traditional role. In ESS, the teacher acts as a guide and as a resource person. The activity is initiated through a discussion, a demonstration, or a question. Students then develop problems and devise solutions. The teacher raises questions but does not answer them. He may verify answers or ask leading questions to provoke thought or challenge data and solutions, but always he is trying to get the students to think and to understand what they are thinking about.

Students also have a different role in ESS science. They are involved participants, not passive receptors. What is learned depends upon what the student does, not on what is read or what the teacher lectures. Students are expected to raise questions, then devise methods of answering them. They develop experiments, collect data, and interpret their findings. Most important, students determine the direction and depth of each ESS unit. General directions are suggested by the ESS teacher's guide, and the teacher steers the investigations, but the curiosity and ability of the students are the final determinants of what is done with each unit. The flexibility of the program allows a curious student to explore any of the numerous problems that arise in the same unit. The students' active participation makes them partners in the learning situation.

The unit *Bones* can be used as an example of how ESS is taught. There are three distinct portions of the unit: (1) "Mystery Bones," (2) "Bone Picture Book," and (3) "Bone Puzzle." To start the unit, the teacher takes one of the mystery bones, shows it to the students, and asks them to describe it and to speculate on it. Where did it come

This group is learning using the ESS unit Bones.

from? What kind of animal? Size? Function? As the various bones are presented, the teacher directs the discussion toward function and then leads the discussion to the bones in the students' bodies. They can feel them, try the different joints, or speculate on their use. In reality, they are developing an awareness of function. Then the "Bone Picture Book" is introduced. The students look at the pictures and discuss them. They learn through their discussions and observations that they can tell a lot about an animal from its skeleton. Without realizing it, they begin doing what a paleontologist does—taking a bone, studying its structure and size, and determining its function. Application of this knowledge is put to work in the third phase—"Bone Puzzles." A complete cat, mink, or rabbit skeleton is given to a small group of students. They are directed to lay it out in a tray in its natural order. There are no pictures or directions for them to follow. They must take each bone, determine its function, and place it in its proper place. This can be a real challenge, but one that fifth or sixth grade children can easily master, if they have worked up to it. The teacher leads the early discussion and then acts as a consultant, but the students do the work. When they finish the unit, they have a good idea of what the skeleton does and how it works.

The procedure of discussion, speculation, experimentation, and application is carried out in virtually all of the units. The teacher acts as a guide and the students investigate. In ESS, learning is an experience shared by the student and the teacher. At this point, you might want to look at one of the teacher's guides. You will get the opportunity to review several when you start on the activities section of this program.

Evaluation of Student Performance

The ESS program is not a traditional type of program; therefore, it cannot be evaluated in the traditional manner. Pencil-and-paper exams on specific subject matter are not advisable since the focus of the program is on problem solving and "doing" science, not content. It is necessary to develop a different approach to student evaluation. Actually, two methods of evaluation are recommended. The first, and most widely used, is the honest appraisal approach, and the second is the behavioral objective approach.

The honest appraisal approach to evaluation is very simple. The teacher considers the student's effort, contributions, and performance then makes a judgment as to what has been learned from the experience. It is a very subjective evaluation, but probably the most honest, if the teacher takes the task seriously.

Many teachers do not feel competent to develop their own criteria and have considered this a weak area of the program. As a result, ESS developed a manual

By participating and observing, this teacher can make a good evaluation of each student's performance.

called *McGraw-Hill Evaluation Program for ESS.*[5] It contains unit and evaluation objectives for each of the fifty-six ESS units and a sample check sheet for the teacher. This manual does help the teacher better understand the objectives of each unit and, for this reason alone should be in her hands. Evaluation is simplified, or at least made more objective, since the teacher does have a definite list of objectives that can be marked "achieved" or "not achieved." This can establish a basis for an evaluation of the student. However, the teacher still must consider the honest appraisal in determining a grade or report because evaluation is not just a number on a test in the ESS program. It is an honest effort to determine what the child has accomplished during the course of the unit.

Implementation

Implementation of ESS is very flexible. The program can be put into a school curriculum all at once or on a piecemeal basis. It does not need a stairstep approach as is recommended with other programs. Sometimes a school will pilot some of the units one year to get the feel of the program, then completely implement the program the next year. Other schools may put in the whole program all at once. Still other

[5]William Aho et al., *McGraw-Hill Evaluation Program for ESS* (St. Louis: Webster/ McGraw-Hill, 1974).

schools will use one or two new units each year until they have developed a program. All work equally well.

One major point needs to be stressed. ESS teaching is not like the traditional methods of teaching, and if it is to be successful the teachers using it must be taught how to use it effectively. More than one new program has failed because the teachers did not know what was expected or what was to be accomplished. Therefore, it is imperative in this program, as in others that you will become familiar with, that some type of teacher training be provided. The publishers do not stress this point, but it is important. Preservice sessions before implementing the program to show the teachers how to use the materials or inservice sessions to answer questions as they use the materials are highly recommended by schools who have implemented the program.

Provisions for Special Education

One of the most recent additions to the ESS program is the *ESS/Special Education Teacher's Guide.*[6] Thirty-one of the ESS units have been selected as appropriate for the child with learning difficulties. The guide was designed primarily for use by teachers of educable mentally retarded (EMR) students, but other teachers also will find it useful.

The units are divided into three categories—perceptual, psychomotor, and other appropriate units—and their formats follow the same sequence. First, the "Audience" is described, explaining for what type of child and grade range the unit is best suited. An "Overview" describes the unit and explains what it can do for the child; then "Objectives" are specified. "Ways of Getting Started" offers suggestions to the teacher to help initiate the unit. This is followed by suggestions for "Keeping It Going." "Other Classroom Tips" provides extra help or insight into classroom management. An "Evaluation Checklist" is provided to help in evaluating the unit, and "Time Required" provides guidelines for scheduling. Finally "Ordering Information" describes the ESS materials necessary for the unit.

Teachers who work with the exceptional child will find this addition to the ESS program very helpful. The flexibility of the ESS program makes it easily adaptable to all children.

Summary

The Elementary Science Study (ESS) is a nonsequential, modular science program based on the assumption that science can be fun as well as informative and accurate. Units in physical science, biological science, earth science, and mathematics are available for grade levels K-6. Schools can choose units appropriate to their science program to develop a science curriculum tailor-made for their school. Teaching and learning are in partnership as the teacher and students work together to determine the direction and depth of the units. Little or no reading is required of the students since learning takes place by doing science, not by reading about it. Each unit consists of a kit of materials and a teacher's guide. The materials are for the students to manipulate, and the guide directs the teacher. As the program is not taught in the traditional manner, teachers do need to learn how to use the material, and this can be done through a preservice workshop. The program is effective, but it does require a flexible, secure, and understanding teacher who is willing to let students learn.

---

[6]David W. Ball, *ESS/Special Education Teacher's Guide* (St. Louis: Webster/McGraw-Hill, 1978).

Before you begin these activities, be sure you have read the background information for the ESS program. Now you should have some understanding of it. In keeping with the philosophy of learning by doing, you are going to be directly involved with some of the materials used in the ESS program.

   The first step will be to review some of the teacher's guides. See your instructor for directions in obtaining these. You can get the feel of the program by reading through the same material that the teacher would in preparing to teach a module. What you have read about the program can be compared with the real materials. Remember, the background information can serve as a guide or provide additional information; so refer to it as you go through these activities. After you have reviewed several guides, you will be given some actual ESS activities to do. This will give you the opportunity to see what kinds of materials and activities can be used. Keep in mind that soon you will be teaching and be alert to any idea or activity you could use in your own classroom. Think about what you see and do and how it can help you be a better teacher.

   As you go through these activities, remember that the content is not as important as how it is being used. Look at the material from the student's point of view as well as from the teacher's. Don't forget to discuss your thoughts with your group and with your instructor. You are responsible for what you learn in this module; the instructor is only a guide and a resource person.

These preservice teachers are familiarizing themselves with the ESS activities.

Activity 1: Review of ESS Teacher Guides

A. The teacher's guide is the key to all ESS modules. It gives information about the program, materials needed, and ideas for teaching. Your first task will be to review some of the actual guides in order to do the following:

1. Familiarize you with the ESS guides by having you examine them
2. Help you better understand the teaching strategy used by having you read the suggestions for teachers found in the guides
3. Give you some ideas about how you can use everyday materials to teach children how to do science by showing you some actual examples of how this can be done

As stated previously, the ESS units are grouped into four subject areas: biological science, physical science, earth science, and mathematics. If you need help determining the subject area or the grade range of any of the guides, refer to the chart on pages 45-47. Try to be selective as you make your choice of guides to review. You might want to look at some from each of the different subject areas or at some that are for different grade ranges. Work with your group to determine the most profitable approach. Each member should review several guides and then meet with the group for mutual sharing. (Your instructor might choose to set a required minimum number to be reviewed.) Remember to look for ideas and activities as well as teaching strategies. (Share these with your group, too.) Ask your instructor to participate in your discussions.

B. Now that you have made your choice of guides to review, note the title, subject area, and grade range of each before you start. As you read through them, look for the following information:

1. How is the information in the guide organized? Compare several guides. Are all of them organized in the same manner? What type of information is available to the teacher? Do you think the information is adequate to help *you* teach the module?
2. Find the activities. Are they written up as you expected them to be? How are they taught? Look for examples of student discovery. How are the students involved in the act of discovery? Could you adapt any of these activities to your classroom?
3. Look at the types of materials used. Could you find the necessary materials if you wanted to teach any part or all of the module?
4. What is the teacher's role? Does the manual describe it? Can you find evidence of the teacher's acting as a guide or a resource person? See if you can determine the student's role.
5. Try to find suggested methods of evaluation. Are they what you expected? What is the teacher *really* trying to find out in an evaluation? How could you use this type of evaluation technique in your classroom?
6. What is the module really trying to teach? Look beyond the title and try to determine the real science skills being taught. Are the science processes, such as observing, classifying, etc., taught in the module?
7. Look closely for any information you can use to help you become a better teacher. Think about how you could adapt some of the ideas, suggestions, and activities to your classroom. Remember, one purpose of this exercise is to familiarize you with the ESS program, but the most important purpose is to give you the opportunity to improve your teaching repertoire.

C. You have looked at some ESS guides and should have some interesting comments to make. Here is the place to do it. Write down your thoughts. These comments should be for your own benefit, not necessarily for your instructor's; so say what you think. Remember, you are not expected to like every program you review. All programs have strengths and weaknesses; look for them.

COMMENTS

D. Meet with your group to discuss the guides reviewed. You might want to invite your instructor to participate in the discussion. Share the ideas you have obtained and the concerns you may have about ESS. If you have any questions, bring them up for discussion. Make some notes on your group discussion.

COMMENTS

√ SELF-CHECK

Did you review several guides, or only one? Each guide is different and one guide can't give you a true picture of the program. Did the guides help you understand how the teacher presents the material? There are no specific "do this" or "do that" directions, only general guidelines. Would it be difficult to obtain ESS materials? It should not be, as almost all of the materials can be found around the school or community. Some materials are specialized and require special purchase. Did you see any of these modules? What ideas can you take with you? How would you like to teach from this program?

COMMENTS

SUMMARY

In this activity, you have reviewed several ESS teacher guides to help you better understand the program. The types of activities used, the kinds of materials used, the teaching strategy, and the student's role were stressed. The objectives of this activity were to help you understand how the program is taught and to develop new insight into your own concept of teaching science.

Activity 2: Activities for ESS

Here are four activities taken from ESS material, one from each of the different subject areas. Do at least two of them. (Your instructor may choose to tell you how many to do.) These activities were selected because they provide interesting involvement using simple, easy-to-obtain materials. In general, each activity represents the type of activities found in its subject area grouping. Because of the individual nature of the modules, no activity can completely represent all of the other ESS activities. However, these activities will help you understand what the student does in the ESS science class. The activities are not taken directly from the program but are composites of several activities found in the teacher's guides and are modified to be appropriate for your needs and your limited time.

Select the activities you are going to do by looking at all of them before choosing. Your group might want to decide on the approach you will take to complete these activities, whether you will work on them as a total group or subdivide the work. Do not try to do them individually because you'll miss out on the group interaction that is stressed in ESS activities. Furthermore, don't be afraid to extend the activities beyond what is required. If you think of

Group interaction is important in doing ESS units.

something interesting, try it. If a "what-would-happen-if" situation occurs, find out. That is what ESS is all about.

As you go through the activities, think about what you read in the background information and in the teacher's guides. Try to combine the three elements to give you an understanding of the program and the philosophy behind it. Think about how a child might react to what you are doing. How would you present the same activity to an elementary student? And, most important, don't forget to try the activities. You will miss out on the fun and the learning experience if you don't get your hands on the materials.

Biological Sciences Activity: Mealworms[7]

This activity is a modification of the upper-grade unit *Behavior of Mealworms*. The most important thing a child can learn from this unit is how to carry out an investigation. See if you can do the same. As you do the structured activities, think of other investigations you would like to make. How would you set up the investigation, and what would you look for?

A. In this activity you will become familiar with mealworms. Mealworms are the larval stage of the grain beetle *(Tenebrio molitor)* usually found around flour mills or in grain warehouses and feed stores and are commonly used as fishing bait. (This should give you a clue as to where to obtain a supply if you wish to use this activity with a class at a later date.) Locate the container of mealworms provided, and you will be ready to start.

Many interesting observations can be made about meal worms using a hand lens.

[7]Adapted from ESS, *Teacher's Guide for Behavior for Mealworms,* © 1966 Education Development Center. Used by permission of the publisher, Webster/McGraw-Hill.

Ode to a Mealworm

by Lora Fleming, Park School, Brookline, Mass.

Pity the poor mealworm
He is not an ideal
worm
In fact, he's not a real
worm
But a bug.
Ugh
And when he sought the bran
He couldn't escape my scan
No matter how hard he ran,
What a bug,
Ugh[8]

B. Mealworms and Their Habitat

1. Observe the mealworms in their natural habitat. Describe their habitat.

2. Where do mealworms seem to prefer to stay?

3. Can you find an example of each of the **three stages of development?** Identify and sketch each.

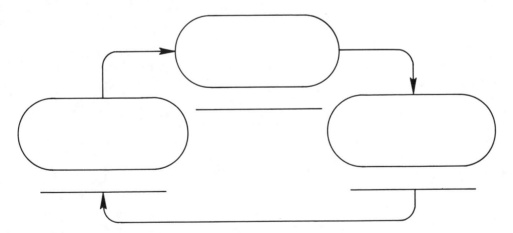

---

[8]From Elementary Science Study, *Teacher's Guide for Behavior of Mealworms* (St. Louis: Webster/McGraw-Hill, 1966), p. 38.

C. Structure of Mealworms

1. Examine the structure of two or three mealworms. Hand lenses are available. *Do not* do anything that might injure the mealworms. Describe the structure of a mealworm; then draw a picture.

D. Learning about Mealworms

1. Investigate at least two of the following questions. Keep in mind that a mealworm has chemical reactors over its entire body, and putting a drop of irritating liquid on a mealworm is somewhat like having it poured into one's mouth or nose. Irritating liquids should be dropped near, but not directly on, the mealworms.
   a. How do mealworms move?
   b. How do mealworms eat?
   c. Can mealworms see?
   d. Can mealworms be made to back up?
   e. Can mealworms follow walls?

   Record the questions selected and your answers to each.

2. What other activities can you suggest for elementary children to investigate mealworms?

√ SELF-CHECK

For the activity:

Did you have any trouble finding examples of each stage of development? Larvae move around and pupae do not. Were you squeamish about handling the organism? Some children (and teachers) are at first but get over it as interest develops. Did you manage to answer two of the questions to your own satisfaction? Did you try anything extra just to satisfy your own curiosity about something that came up during your investigation? You might want to refer to the *Teacher's Guide for Behavior of Mealworms* if it is available.

Implications for teaching:

How would you teach this module to your class? It is an easy, inexpensive way to have living organisms in your classroom. What other simple organisms might you use? There are many varieties available, such as earthworms, pill bugs *(Isopods),* or ants. Could you make up your own ESS activity? A good idea is versatile and can be adapted to almost any situation.

COMMENTS

SUMMARY

A simple, easily maintained organism is used to teach children how to experiment with habitat, environment, and climatic conditions necessary for optimum development of an organism. This unit is inexpensive and requires very little equipment. Many teachers are using earthworms, rather than mealworms, as they are more readily found by children. Other related ESS units are *Brine Shrimp, Butterflies, Crayfish, Eggs and Tadpoles,* and *Mosquitoes.* You might want to examine some of these manuals if they are available.

Physical Science Activity: Colored Solutions

Colored Solutions is a physical science activity best suited for grades three to six. Density and layering of liquids are used to help children learn how to answer questions using both physical materials and their own investigations. Two simple activities are provided here to introduce you to the unit.[9] Remember, these are basic, introductory exercises, but the students in your classroom could do a lot more with this unit. Can you?

---

[9]Adapted from ESS, *Teacher's Guide for Colored Solutions,* © 1968 Education Development Center. Used by permission of the publisher, Webster/McGraw-Hill.

Using colored solutions, students learn about the density of liquids.

A. Dropping, Mixing, Watching

1. Fill a plastic cup with water.
2. Put one drop of food coloring in the water. Do not stir. Now, describe and draw what you see.

3. Try adding different colored drops, one at a time. Describe what you see.
4. Describe what happens if you stir the water after adding one color. After adding two or more colors.
5. Would there be any difference in the results if you used warm water or cold water rather than room-temperature tap water? Find out.

B. Layering

1. Four containers of liquid are used in this activity. They are numbered and colored. Three of the solutions are salt water of varying densities, and one is tap water. The salt solutions are made by adding 1 cup, 2/3 cup, and 1/3 cup of salt, respectively, to three one-quart containers of water. A fourth one-quart container has no salt added. Three of the containers are then

colored with green, red, and blue food coloring; the fourth is left clear. Only your instructor knows which solution has no salt and what combinations of salt and water are in the other three colored solutions. You are going to find out for yourself.

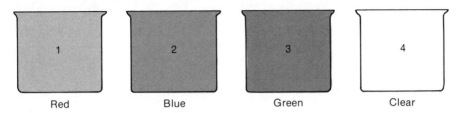

| Red | Blue | Green | Clear |

2. By *carefully* placing one solution on top or underneath another, you can find out which is denser. A dropper or a straw can be used to add the color slowly and carefully.

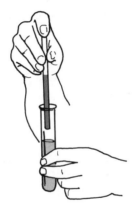

Special Hint: Use a small diameter vial or test tube rather than a large container like a plastic cup. It works better and is easier to see. Try using the red and blue solutions. Which is the more dense of the two? How do you know? Draw your results.

3. Try floating one color on another to make a "parfait." See if you can layer all four of the liquids. Which color is the most dense? Which is the least dense?

4. Sketch a picture of your parfait and label the colors. You should now know which colored solution is what salt combination; so indicate that on your drawing.

√ SELF-CHECK

For the activity:

Were you surprised by the patterns made by a single drop of food coloring placed on the water? Pretty, isn't it? Did you wonder about the motion of the color in the still water? How could you account for this? You might want to think about molecular motion. Did your colored solutions layer, or did they mix? If you mixed them, try again, *carefully.* You might want to look at the *Colored Solutions* guide it if is available.

Implications for teaching:

Do you think that children would like this activity? Did you? How expensive would it be to use this activity? Salt and food coloring can be purchased at any grocery store. What else can you think of to do with this activity? Did you try anything not in the directions? Why not? You have time, and the equipment is available. Wouldn't you want your students to answer some of their own questions if you were the teacher?

COMMENTS

SUMMARY

Food coloring, salt, and water are used to teach children how to investigate density and molecular motion. The students learn how to experiment in order to answer questions that develop as they do science activities. Related units are *Drops, Streams and Containers, Kitchen Physics, Sink or Float,* and *Water Flow.* You might look at some of these guides if they are available.

Earth Science Activity: Mapping

In the study of earth science, it is important that the student have some knowledge of physical geography. One important aspect of physical geography is knowing how to read and construct a map. Even though the study of maps and map skills is usually encountered in social studies, it can also be taught, with relevance, through the science program. ESS presents this material through a unit entitled *Mapping*. Several ideas are presented, such as playing games to develop map skills, constructing maps or models, and following and giving directions. This activity concentrates on following and giving directions.[10]

A. Following Directions

1. Obtain a set of directions from your instructor. This should be a handout prepared by your instructor with directions to one or more places in the building or to objects within the room.
2. Follow the directions. Where do they lead you?

3. Did you have any trouble following the directions? What changes would you make in the directions so that they are easier to follow?

B. Giving Directions

1. Select one of the following locations:
   a. Your dormitory or apartment
   b. Your favorite local restaurant
   c. Registrar's office
   d. University bookstore

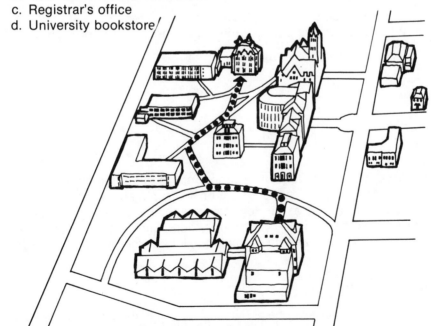

---

[10]Adapted from ESS, *Teacher's Guide to Mapping*, © 1968, 1971 Education Development Center. Used by permission of the publisher, Webster/McGraw-Hill.

2. Write out a set of directions that would get you from your classroom to there. Pretend that someone who does not know where the location is must follow your directions. Use as many steps as you feel are necessary. Location:

Directions:

3. Draw a map showing how to get from your classroom to the point selected.

4. Have other members of your group read your directions and look at your map. What suggestions did they make to improve either or both?

√ SELF-CHECK

For the activity:

Did you have any trouble following directions? Is this something you need to work on? Can you give good directions? Think about how giving and following directions are a part of map usage. Be sure to compare your directions with other members of your group. Can you follow their directions? If the guide for *Mapping* is available, look through it.

Implications for teaching:

You can combine your science activities with other subject areas. Here is an example of science and social studies. Can you combine art, music, mathematics, and physical education, as well as reading, with your science program? A good teacher can. Something else to think about: Can you give clear instructions to your students? Teachers who give unclear instructions cause failures in assignments and frustrating situations in the classroom.

COMMENTS

SUMMARY

Mapping helps students understand the physical relationships found in the earth's surface. In this unit students learn how to use directions and graphic representation to better group the concepts of location and symbols. Related ESS units are *Stream Tables, Match and Measure, Daytime Astronomy, Light and Shadow,* and *Geo Blocks.* You might be interested in reviewing some of these guides.

Mathematics Activity: Peas and Particles

A word of explanation about ESS mathematics modules is needed before you begin *Peas and Particles.* Spatial relationships, measuring, attributes, and the use of mathematics are the subjects of these modules, not numerical computation. You might want to refer again to the chart on pages 45-47 to see what titles are included in this category.

This teacher is introducing a new science approach to his students.

This activity is one in which children deal informally with estimation and large numbers in ways that may be new to them.[11] They answer questions— How many? How big? How far away?—not with worksheet or arithmetic test precision, but as we tend to answer questions ordinarily—with wild estimates and educated guesses. For this activity, all you need is something to estimate such as a pint of rice or peas, a container of marbles, a quart of macaroni, or a quart of beans. Your imagination is your guide.

[11]Adapted from ESS, *Teacher's Guide for Peas and Particles,* © 1966, 1969 Education Development Center. Used by permission of the publisher, Webster/McGraw-Hill.

A. Estimating a Handful of Peas

1. You will need to work with at least one other person in this activity. One of you should take a handful of peas from the container of peas.
2. Observe the handful of peas for a few seconds. Now, both of you are to make an estimate as to the number of peas in the handful. Your guess. Your partner's guess.
3. Count the peas and compare the actual count to the estimates. Actual number in the handful. Who was closer, you or your partner? By how many?
4. Try it again, letting the other person get the handful this time. Your estimate. Your partner's estimate. Actual count.
5. Did you come closer the second time? Why or why not?

B. Estimating a Jarful

1. With your partner, select one of the jars of objects provided for this activity. Which jar did you select?

2. Each of you should make a quick guess, without taking the objects out of the jar, as to how many objects are in the jar. Your guess. Your partner's guess.
3. Now try some serious estimating to check your hasty guess. You may want to actually count the objects one by one, but don't do that just yet. Try to develop some more imaginative methods of estimating to check your guess. Devise some counting strategies that involve manipulating the objects. You are to describe your strategies and make your estimate based on each strategy.

Strategy 1:

Estimate:

Strategy 2:

Estimate

Strategy 3:

Estimate

4. If you feel like counting the objects to check your guess and your estimating strategies, you may do so now.

SELF-CHECK √

For the activity:

Did you find that your ability to estimate improved with practice? What caused the improvement? Developing a point of reference, such as how many peas in a handful, gives you something concrete to base your estimate on. Did you have to resort to counting, one by one, to confirm your estimates, or were you willing to trust your estimates? If you had a bushel of beans, would you trust a carefully planned estimate, or count them one by one? If the guide is available, you might refer to it at this time.

Implications for teaching:

You can use this activity to introduce either estimating or large numbers to a class. How hard would it be for you to obtain objects to estimate? One teacher actually used a bushel of beans to initiate the activity. Can you think of a better attention getter than a contest to guess the number of beans in the bushel container? What could you do with this activity?

COMMENTS

SUMMARY

Many times a close guess is sufficient or even more desirable than an actual count. With this unit, children are taught to make estimates of large quantites of small objects, using household items. They can then expand into other areas of measurement. Large numbers are also considered and become more real when students can actually see the quantity represented by the numbers. There are no related units available.

Summary: ESS

This concludes the *Elementary Science Study* portion of Part 2-1. In the background information, you read about the program and learned how it was developed, what teaching strategy is employed, and the physical make-up of the kits. In doing the activities, you were given the opportunity to have hands-on experiences similar to those an elementary student might have. Throughout the readings and activities, it has been stressed that you should think about how you might use what you are learning in your own classroom. You are learning about ESS, not because it is the program you will be teaching, but because of the unique teaching strategy it presents.

Now that you have completed this section of Part 2-1, you should be able to do the following:

1. Identify and describe the materials developed by ESS
2. Identify and describe the scope, sequence, and teaching strategy of ESS

If you think you can perform these objectives, you are ready for the final seminar and instructor check. However, if you think you need more information or additional help, ask your instructor to help you.

FINAL SEMINAR

Go back over the ESS activities and review the questions and comments you made. Discuss your reactions with a small group of your classmates. Be sure to discuss any specific questions you may have. You may also share your opinions of the program. *Be sure to invite your instructor to share in this final group seminar.*

NOTES

BIBLIOGRAPHY

Aho, William et al. *McGraw-Hill Evaluation Program for ESS.* St. Louis: Webster/McGraw-Hill, 1974.

Ball, David W. *ESS/Special Education Teachers Guide,* St. Louis: Webster/McGraw-Hill, 1978.

Bruner, Jerome. "The Act of Discovery." *Harvard Educational Review* 31 (Winter 1961): 21-32.

Elementary Science Study. *A Working Guide to the Elementary Science Study.* Newton, Mass.: Education Development Center, 1971.

SUGGESTED READING

Aho, William et al. *McGraw-Hill Evaluation Program for ESS.* St. Louis: Webster/McGraw-Hill, 1974.

Ball, David W. *ESS/Special Education Teacher's Guide.* St. Louis: Webster/McGraw-Hill, 1978.

Carin, Arthur A., and Sund, Robert B. *Teaching Science through Discovery.* 3d ed. Columbus, Ohio: Charles E. Merrill Publishing Co., 1975. Pp. 46-51.

Elementary Science Study. *A Working Guide to the Elementary Science Study.* Newton, Mass: Education Development Center, 1971.

Gega, Peter C. *Science in Elementary Education.* 2d ed. New York: John Wiley & Sons, 1970. Pp. 576-601.

Renner, John W., and Ragan, William B. *Teaching Science in the Elementary School.* New York: Harper & Row, Publishers, 1968. Pp. 287-94.

Science Curriculum Improvement Study

Background Information

Authors:	SCIS	The Science Curriculum Improvement Study Robert Karplus, director University of California, Berkeley, California
	SCIIS	Herbert Thier, Robert Karplus, Chester Lawson, Robert Knott, and Marshall Montgomery (all were members of the original SCIS team)
Publisher:		Rand McNally & Co. P.O. Box 7600 Chicago, Illinois 60680
Author:	SCIS II	Lester Paldy, Leonard Amburgay, Francis Collea, Richard Cooper, Donald Maxwell, and Joseph Riley
Publisher:		American Science and Engineering, Inc. (AS & E) Education Division 20 Overland Street Boston, Massachusetts 02215

At present, there are three versions of the *Science Curriculum Improvement Study* on the market. In 1978, *SCIIS* and SCIS II were published to update and replace the original. The original SCIS program will be discussed first, in detail, to give the background of the project and to describe the basic program. A section on

This student is learning science from hands-on experience.

each of the revisions will follow, explaining their similarities to and differences from the original version.

History

The *Science Curriculum Improvement Study* officially came into existence in 1962, but its beginnings go back to 1957. The central figure in the development of the SCIS program has been Dr. Robert Karplus, a theoretical physicist at the University of California at Berkeley. In 1957, Karplus' children entered the public schools, and he soon became interested in working with their science classes. At the same time, the National Science Foundation (NSF) showed a willingness to support the development of elementary level science programs. The *University of California—Elementary School Science Project* (UC—ESSP) was born. As work continued, a new program began to take shape, and in 1962 it was separated from the parent (UC—ESSP) project. A new name, *Science Curriculum Improvement Study* (SCIS), was adopted.

Karplus worked with the *Elementary Science Study* (ESS) program and with the *Minnesota Mathematics and Science Teaching Group* (MINNEMAST), gaining new ideas and insights that he would later incorporate into his own SCIS program. In 1963, Karplus began to choose the staff that would write, develop, and test the new SCIS program. This staff was composed of university professors, public school curriculum specialists, and classroom teachers. The classroom teachers were considered the indispensable backbone of program development.

Much work went into the development of an SCIS unit. The *SCIS Final Report* describes fifteen steps that were followed in the preparation of each unit. The three key teaching steps are underlined.

1. Brainstorming to create ideas for activities and investigations compatible with the conceptual organization of the unit.
2. Laboratory investigation by project staff of phenomena and devices proposed for inclusion in the unit. Redesign or elimination of materials that appear too difficult to manipulate, too hard to observe, or too unreliable using the equipment and facilities available in elementary school classrooms.
3. <u>Exploratory teaching</u> based on a teaching outline and discussion within the development team.
4. Preparation of a written draft for the trial edition teacher's guide and student manual. (Sometimes this step accompanied the exploratory teaching.)
5. In-house review, revision, and editing of the trial edition manuscript. Design and manufacture of trial edition equipment for about three hundred students.
6. <u>Trial teaching</u> in the Berkeley area by regular classroom teachers.
7. Evaluation of the trail teaching based on observer reports, teacher feedback, used student manuals, and testing of students.
8. Redesign of the conceptual organization, teaching activities, student manual, and/or equipment, accompanied by exploratory teaching, and leading to preparation of a new manuscript for the preliminary edition. (Some units were published in a revised trial edition and were cycled through steps 5, 6, 7, and 8 a second time.)
9. In-house review, revision, and editing of the preliminary edition manuscript.
10. Publisher's editing of the manuscript and apparatus maker's design of commercial equipment for the preliminary edition.

11. <u>Field test</u> of the preliminary edition in the five national SCIS trial centers.
12. Evaluation of the field test based on observer reports, recommendations of trial center coordinators, suggestions from trial center teachers, review of used student manuals, and testing of students. (The upper-grade units, which were not published in a preliminary edition, were field-tested in the trial edition.)
13. Redesign of the conceptual organization, teaching activities, student manual, and/or equipment, accompanied by exploratory teaching, and leading to a new manuscript for the final edition.
14. In-house review, revision, and editing of the final-edition manuscript.
15. Publisher's editing of the manuscript and apparatus maker's design of commercial equipment for the final edition.[12]

All units, in final form, were available for distribution in the fall of 1971, with the completion of the *Energy Sources* unit for fifth grade. A kindergarten unit has since been developed. The project team has been disbanded, but several projects having roots in the SCIS programs have been developed or are in progress. Probably the most interesting is the *Science Activities for the Visually Impaired* (SAVI) program in which science activities are being adapted for use by visually impaired students. All copyrights for the SCIS program passed from the copyright holders (The University of California) to the public domain on January 1, 1978.

Conceptual Background

The conceptual framework for the SCIS program is basically Piagetian as adapted and applied to American education by Bruner and Stendler. These authors stress that even though children go through sequential stages, they do not suddenly pass from one stage to another. For instance, they may be able to use conversation logic but may not be able to reason about abstraction without concrete analogies. Therefore, a science program for the elementary grades should provide a diverse program with emphasis on concrete experiences.

However, providing concrete experiences is not sufficient to insure learning. The experiences must be presented in such a way as to build a conceptual framework that children can use with abstractions. In other words, through the concrete experiences, children collect data to interpret and act upon. As they learn to abstract, they are able to assimilate data collected by others. When this assimilation process occurs, it can be said that the children have developed "scientific literacy". This is the goal of the SCIS program.

Scientific literacy results from basic knowledge, investigative experiences, and curiosity. In the SCIS program, basic knowledge is gained through four major scientific concepts: matter, energy, living organisms, and ecosystems.

Matter is tangible. It can be perceived by the senses, and it has properties. Students become aware that matter can be identified by its properties and can change and interact with other matter.

Energy is not so tangible, but it can bring about change. Energy sources are studied as well as are energy receivers, culminating in the interaction that is a result of energy transfer.

---

[12]Reprinted with permission from the *SCIS Final Report,* published by the Science Curriculum Improvement Study. Copyright 1976 by the Regents of the University of California.

Living organisms are plants and animals that are composed of matter and are able to use energy for their own benefit. This represents a combination of matter and energy.

Ecosystems are the interrelationships of all of the living organisms. This concept considers the diversity of organisms as they interrelate to each other. Every living organism is dependent upon other organisms (for example, the carbon dioxide cycle).

Investigative experiences are an integral part of the SCIS program. The theme might easily be "Don't tell me, I'll find out." Activities are the core of the program. Students are given problems to solve and the materials to solve them with. Furthermore, they are encouraged to investigate any new problems that may arise from the original investigation. Students ask themselves, "What would happen if . . . ," then try to answer the questions.

SCIS also recognizes the importance of teaching the science processes of observing, describing, comparing, classifying, measuring, interpreting evidence, and experimenting. Four concepts—property, reference frame, system, and model—are used to help the student understand these processes.

Program Description

The SCIS program is designed as a complete K-6 science program. It does not need additional texts, references, or other supplementary material. This does not mean that the teacher cannot use supplementary material if he so chooses.

There are thirteen units in the SCIS program. One unit, *Beginnings,* is an introduction to science for kindergarten. The remaining twelve units are divided into two groups, life science and physical science. There is one unit of each group for grades one to six. To best understand what is taught in each unit, here is information from an SCIS promotional brochure by Rand McNally & Co.[13]

Level 1—Overview

The first-year units have certain common objectives: to sharpen children's powers of observation, discrimination, and accurate description. The objectives are accomplished as children care for aquatic plants and animals, raise seedlings, and investigate the properties of a broad range of nonliving objects. The units can be taught in either order.

Life Science Concepts		Physical Science Concepts	
ORGANISMS		*MATERIAL OBJECTS*	
organism	habitat	object	serial ordering
birth	food web	property	change
death	detritus	material	evidence

Level 2—Overview

In both second-year units the theme is change, observed as evidence of interaction or by the development of an animal or plant. The two units therefore require children to add the mental process of interpreting evidence to the observational skills they developed the first year. In their laboratory work children use magnets, batteries, wires, various chemicals, photographic paper, pulleys, electric motors, seeds, mealworms, frog eggs, and fruit flies. The units can be taught in either order, or simultaneously.

---

[13]*Science Curriculum Improvement Study* (Promotional Brochure R 10/71 90281) (Chicago: Rand McNally & Co., 1971).

Life Science Concepts		Physical Science Concepts
LIFE CYCLES		*INTERACTION & SYSTEMS*
growth	generation	interaction
development	biotic potential	evidence of interaction
life cycle	plant & animal	system
genetic identity	metamorphosis	interaction at a distance

Level 3—Overview

Children observe and experiment with increasingly complex phenomena as they build on the first two years of the SCIS program and move toward understanding the energy, matter, and ecosystem concepts. In the physical science unit children experiment with matter in solid, liquid, and gaseous forms, and make and analyze measurements. In the life science unit the children observe the interactions of various organisms within a community of plants and animals and consider the interdependence of individuals and populations within the community.

Life Science Concepts		Physical Science Concepts	
POPULATIONS		*SUBSYSTEMS & VARIABLES*	
population	plant eater	subsystem	solution
predator	animal eater	histogram	variable
prey	food chain	evaporation	
community	food web		

Level 4—Overview

In the life science unit, children consider for the first time some of the physical conditions that shape an organism's environment. These investigations make use of the measurement skills and scientific background developed in the physical and life science units during the first three years. The physical science unit introduces techniques for dealing with spatial relationships of stationary and moving objects.

Life Science Concepts	Physical Science Concepts
ENVIRONMENTS	*RELATIVE POSITION & MOTION*
environment	reference object
environmental factor	relative position
range	relative motion
optimum range	rectangular coordinates
	polar coordinates

Level 5—Overview

The conceptual development of the SCIS program continues as examples of energy transfer are introduced in the physical science unit and of food transfer in the life science unit. Children apply the systems concept, the identification of variables, and the interpretation of data with which they have become familiar during the earlier years of the SCIS program.

Life Science Concepts		Physical Science Concepts
COMMUNITIES		*ENERGY SOURCES*
producer	community	energy transfer
consumer	food transfer	energy chain
decomposer	raw materials	energy source
photosynthesis		energy receiver

Level 6—Overview

The last year of the SCIS program contains both a climax and a new beginning. The life science unit integrates all the preceding units in both physical and life sciences as children investigate the exchange of matter and energy between organisms and their environment. The physical science unit introduces the concept of the scientific model and thereby opens a new level of data interpretation and hypothesis making. At the same time, the children relate matter and energy to electrical phenomena, acquiring a basis for their later understanding of the electrical nature of all matter.

Life Science Concepts	Physical Science Concepts
ECOSYSTEMS	*MODELS: ELECTRIC & MAGNETIC*
	INTERACTIONS
ecosystem	scientific model
water cycle	electricity
oxygen-carbon dioxide cycle	magnetic field
pollutant	
food-mineral cycle	

The key word to the SCIS program is *interaction*. As objects or living things interact with each other, change occurs. This change is natural and predictable. SCIS teaches children to look for change as evidence of interaction among objects.

The SCIS program is sequential. Each life science unit builds on prior units, as does each physical science unit. Students must have the information obtained at each level before they proceed to the next level. This brings up the question of transfer students. How do they fit into the program? Easily, says SCIS. The teacher works with the student, just as she would with any student who has been involved in a different program, and the student can quickly grasp the concepts and work patterns and move right into the class activities.

The materials for teaching SCIS science are in kit form. Everything needed to teach the unit is included, even live organisms and supermarket items. In each kit there are materials for thirty-two students including student manuals and a teacher's guide. A complete list of materials is included in the teacher's guide. SCIS kits can also be shared by several teachers, a practice sometimes necessary to cut back on costs. Refill kits are available so that each teacher has his own set of consumable items but shares the basic kit.

The cost of implementing the SCIS program is comparable to any other program. Two kits for each teacher are obviously recommended, but kit sharing is an effective way to reduce costs.

Materials for six SCIS kits.

Teaching Strategy

SCIS employs a unique teaching strategy called *exploration—invention—discovery*. It is a strategy well worth knowing and using. The teacher and the students assume different roles in each phase of this learning cycle.

Exploration consists of getting materials into the hands of the students, arousing their curiosity, and allowing them to find out what they have and what they can do with it. This is an often neglected, but vital, part of teaching. The students learn through their own spontaneous activities and experiments. The teacher's role in exploration is that of a guide who encourages the students but uses a minimum of specific instruction. The students are the active partners in this phase of the learning cycle. They handle the equipment, try out ideas, share ideas and results with each other, and develop new ideas of things to do. One thing to remember, though—it takes time for a student to explore. It is a process that cannot be hurried.

Invention is the structured teaching phase of the learning cycle. After the students have made some observations and discoveries in the exploration phase, they need help to develop the new concepts necessary for them to understand what they have learned. The teacher, acting as a resource and a guide, brings the class together to discuss their explorations and to "invent" the concept that they are working with. The teacher may ask questions, draw pictures on the chalkboard, lead discussions, or give verbal explanations during this phase. The student assumes the traditional role of learner but also participates in discussions by raising questions and suggesting answers. The goal is to use what was learned in the exploration phase to "invent" a concept previously unknown to student.

Discovery can be called the application phase of the learning cycle. Students participate in activities which allow them to discover new applications for the concepts that they have just invented. The teacher may provide the applications or may depend on the students to furnish their own. The teacher again acts as a guide or resource person as needed. The students actively participate as learners by doing activities that reinforce the original concept.

As part of "Clues for Teachers" each SCIS teacher's guide explains these three phases of the learning cycle and describes where each can be found in the guide. Look for this when you review the teacher's guides in the activities section.

The teacher's guide tells you how to teach SCIS. Reading through the guide is, of course, the first step. Here you find a program overview, helpful hints on teaching called "Clues for Teachers," a description of the kit listing all of the materials included, and the actual teaching unit, divided into chapters. For each lesson, the guide gives you an overview of the lesson, materials needed and their location in the kit, advance preparation needed, and teaching suggestions. Follow the guide, and you will also follow the exploration—invention—discovery learning sequence.

Remember, take time to let your students *explore* before trying to *invent* the concepts. Then, let them *discover* new ways to use their newfound knowledge. This way, the students can gain investigative experience, learn basic knowledge, and keep curiosity alive, thus fulfilling the SCIS requirements for scientific literacy.

Evaluation of Student Performance

SCIS is not a traditional, content-oriented program and cannot be evaluated as such. Written quizzes on factual material are not practical because there is very little factual information presented. The students work toward the goal of scientific literacy, and the teachers must be the judge of the progress being made toward that goal.

Evaluation in SCIS is considered to be a very important on-going process upon which the teacher bases his teaching. Constant evaluation of each student's progress, based on a variety of feedback sources, allows the teacher to determine what each student is doing and what her needs are at the time. The teacher can then guide the student into new discoveries that will bring her closer to scientific literacy.

Feedback is obtained in many ways. There are suggested feedback activities in each SCIS chapter to help the teacher determine what progress is being made toward the chapter goals. One of the easiest methods to obtain feedback is simple observation of the students to find out about their work habits, attitudes, and classroom participation. Written work on lab sheets and in the student manual also provide clues for the teacher.

SCIS also encourages the use of questioning as a means of evaluating pupil progress. Convergent and divergent questions are recommended. Convergent questions tend to produce factual information and are effective when evaluating content knowledge. Divergent questions can be described as thinking questions. Various answers are possible, and each answer reflects the knowledge and interpretation of the student giving it.

Classroom teachers seemed to be happy with this more informal approach to evaluation, but school administrators were not. In response to the requests of the administrators, the SCIS project team developed a more formal evaluation program called *Evaluation Supplements*.[14] These areas are covered in this program: (1) student perception of the classroom environment, (2) process content objectives, and (3) attitudes in science. Evaluation activities and charts for recording the evaluative outcomes are provided. Two other areas, general intellectual development and teacher self-evaluation, were also considered important. But after field testing, these two areas were removed from the *Evaluation Supplements* and included in the *SCIS Teacher's Handbook*.[15] Most of the teachers who use these supplements do not feel they add very much to their regular evaluation, and the increased record keeping takes up time that could best be used for teaching. The *Evaluation Supplements* fulfill an administrative need but are not included extensively when SCIS is implemented.

In summation, the SCIS program is evaluated by teachers as they teach and by students as they learn. Conscious and subconscious observations lead teachers to adjust their approach to fit the needs of their students, and students respond based on their own perception of their needs and goals.

---

[14] *Evaluation Supplements* (Berkeley: Science Curriculum Improvement Study, University of California, 1971-1975.

[15] Robert Karplus and Herbert Thier, *SCIS Teacher's Handbook* (Berkeley: Science Curriculum Improvement Study, University of California, 1974).

Implementation

The SCIS program can be put into effect all at once or in a stair-step manner. The stair-step approach allows a school system to implement the program over a period of two or three years. It also allows for the initial cost to be spread over a period of time. On the other hand, a school system may choose to implement the program all at once to get everyone involved in the program at one time. This way, no one is left out or has to wait for the new equipment. Variations of these implementation procedures can be devised to fit the specific needs of a school.

Some special considerations are required in the implementation of the SCIS program, but probably no more than should be made in the implementation of any new program. As mentioned in the ESS section, too often, programs such as SCIS, SAPA, and ESS fail because the teachers are not properly prepared. They are simply given the materials and told to teach them without knowing what to do with the program, what to expect from it, or how to start. For this reason, preservice and inservice programs must be provided for the teachers who will be using SCIS. Help can be obtained from the publisher or from consultants recommended by the publisher.

Summary

The SCIS program is a sequential program for grades one through six. A kindergarten program is also available. The program was developed by Robert Karplus under a National Science Foundation grant and is marketed by Rand McNally & Company. Scientific literacy, defined as a functional understanding of scientific concepts, is the program goal. (The teacher uses the strategy of exploration—invention—discovery to develop scientific literacy.) Two units per grade level, one on life science and the other on physical science, are taught, using the concept of interaction. Students are encouraged to learn through experimentation and to maintain their curiosity. There is very little material to be read by the student, but the teacher can encourage supplementary reading as necessary. The kit contains everything needed to teach the program including supermarket items, consumable materials, live organisms, and reusable hardware. Refills are available for either kit sharing or for the beginning of a new year. Teachers need some training if they are to get the most out of the program.

The Rand McNally SCIIS Program

SCIIS was introduced in 1978 by Rand McNally & Company to replace the original *Science Curriculum Improvement Study* materials that they had published since 1970. All of the authors of the new program were members of the original SCIS project team. The title *SCIIS* was chosen to indicate that even though the program was a revision of SCIS, the revisions were *internal* and not merely external appendages; hence the inclusion of the extra *I* (or Roman numeral *II*, if you prefer) in the title. *SCIIS* is not an acronym but is the title of the program.

SCIIS is basically the same program as the original SCIS but with modifications that reflect the suggestions of the SCIS users. Almost all of the information concerning the conceptual background, program description, teaching strategy, evaluation of student performance, and implementation given for SCIS can apply equally to *SCIIS*. Moreover, Rand McNally offers a conversion kit for changing SCIS kits into the new *SCIIS* program.

The major changes found in the *SCIIS* program are as follows:

1. Earth science has been included in the program. The two major sequences have been renamed life/earth science and physical/earth science. New material has been added emphasizing the student's physical relation to the earth and the ecological nature of the earth.
2. Some activities have been changed. These new activities may incorporate hardier organisms than before, introduce new material not covered before, or be the result of suggestions of a better way to teach existing material.
3. The level five physical/earth science unit has been expanded to include solar energy.
4. The level five life/earth science unit adds *reproduction, the pyramid of numbers,* and *competition* to the original concepts for this level. Competition for food is stressed.
5. The level six physical/earth science unit undergoes a major change, becoming *Scientific Theories.* Color, electricity, magnetic field, and light rays are covered.
6. The teacher's guide has been redesigned. It is larger and incorporates many more sketches to help the teacher understand the text material or equipment being used.
7. An appendix in the teacher's guide contains evaluation procedures for each chapter. This is completely new material.
8. *Extending Your Experience* (EYE) cards are a new addition to each unit. These cards provide activities to supplement the regular activities. Designed to be used by the individual student, the cards can provide enrichment, remediation, or extension of the basic topic. The teacher's guide provides suggestions for their use.
9. The design and packaging of the program have been modified for easier storage and handling.
10. Supply sources, delivery systems, and teacher services have been improved.

In summation, *SCIIS* is an improved version of the basic SCIS program, according to Rand McNally. New material, new packaging, and improved support add to a solid core program.

The American Science and Engineering SCIS II Program

The copyright for the federally funded SCIS program became public domain in 1978. American Science and Engineering (AS&E) had been the exclusive designer and producer of all of the materials (except printed material) used in the SCIS programs during the copyright years. Because of their experience with the SCIS program, it was only natural that they would start publishing and marketing SCIS material when the copyright expired. The original SCIS program is available from AS&E but is being replaced by the newer SCIS II program.

SCIS II uses the original SCIS program as a foundation, retaining the basic *conceptual background, program description, teaching strategy, evaluation of student performance,* and *implementation* that describe the SCIS program. Its authors are respected science educators who have had extensive experience in training teachers to use SCIS as well as in implementing and managing SCIS in their own classrooms. Their experiences led them to turn SCIS into a more manageable program for the classroom teacher. A conversion kit (available from AS&E) allows present SCIS users to change to the newer SCIS II program with a minimum of expense.

The major changes or innovations in the SCIS II program are as follows:

1. The teacher's guide has been changed to emphasize classroom management. This change gives teachers help where they need it most.
2. New content has been introduced. The addition of earth science in the fourth and sixth levels is the most noticeable change. One or more additional concepts are introduced in each unit of the program.
3. The title of the fourth level physical science unit has been changed to *Measurement, Motion, and Change.* The concepts reference frame, change, distance, direction, and measurement were added to reflect the inclusion of earth science at this level.
4. The title of the sixth level physical science unit has been changed to *Modeling Systems.* The concepts listed for this unit are electricity, model, magnetism, circuit, electrical energy, air temperature, barometric pressure, and atmosphere.
5. Activity cards are provided for use by individuals or small groups. These cards provide additional activities for enrichment and relate the science lesson to math and language arts.
6. The student manuals have been replaced by duplicating master booklets. This cuts cost and simplifies classroom management of the materials.
7. The SCIS terms *exploration—invention—discovery,* used in describing the learning cycle, have been replaced in SCIS II by the terms exploration—concept—application. The learning cycle is still the same, but different terms describe the same step.

In summation, the AS&E SCIS II program is the basic SCIS program modified to include earth science and to make it more manageable for teachers.

Now that you have read the background information for SCIS, you are ready to do some activities to see firsthand how the program works. You will review some of the SCIS teacher's guides to give you a "feel" for the program before going on to some of the actual activities. In the guides, you will find teaching suggestions, ideas for activities, and lists of materials. The SCIS activities presented are from both the life science and physical science units.

Remember to look for ideas and suggestions that you can adopt for your own teaching. SCIS may or may not be the science program in the school where you will teach, but that is not important. What the program can teach you is important. Can you see how the SCIS teaching strategy works? How could you use it? How do you think children would react to some of these activities? Is it anything like the science program by which you were taught? There are many ways to teach, and a good teacher will use as many ways as possible to reach all of the students.

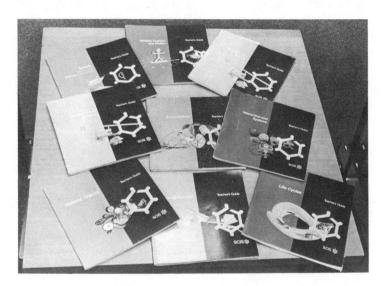

Nine teacher's guides for SCIS activities.

Activity 1: Review of SCIS Teacher's Guides

A. Your first hands-on activity will be to review some of the SCIS teacher's guides. There are four reasons for this activity:

1. To familiarize you with the SCIS teacher's guides by having you examine them
2. To help you understand how the program is structured by having you examine different guides to see the life science and physical science sequences
3. To help you better understand the teaching strategy by having you read the directions and teaching suggestions found in the guides,

4. To give you new ideas about teaching and the types of materials that can be used by having you examine the actual suggestions made by SCIS

You might want to refer to the grade-level overviews on pages 73-75 to refresh your memory about the units and to help you determine which guides you would like to review. You should select guides that will give you a good cross-section view of the program. Your instructor might want to set a minimum number of guides to be reviewed, but you should examine as many as you feel are necessary for you to understand the program.

Review the guides alone or in small groups. If you review alone, plan to meet with your group for information sharing. You might want to invite your instructor to share in some of your group discussions. Don't underestimate the value of group interaction, either as a student involved in it or as a teacher using it, as a teaching device.

B. Make a note of the titles, subject areas, and grade levels of the guides you have selected. Look for the following information as you read through them:

1. Find the section entitled "The SCIS Conceptual Framework." Does this section help you understand that interaction is the core of SCIS? How are process-oriented and scientific concepts used in the program?
2. Read the "Program Overview" to get an overall picture of SCIS. Does this clarify the scope and sequence for you?
3. "Clues for Teachers" should give you some information about teaching strategy. Could you use some of these suggestions in your daily teaching routine? Note the information on the exploration—invention—discovery learning cycle and on the use of questioning in the classroom. These suggestions can be used in any classroom.
4. Spend a little time reading "Design and Use of the Kit." This section will tell you what is in the kit, how it is packaged, and what you will need to supply. The life science guides also offer a schedule of activities. What is the purpose of this schedule?
5. Find the "Activities" section. How is it organized? What type of information is included? Can you think of anything not included that should be? Could you adapt any of these activities to your classroom? Which one is your favorite?
6. Can you find examples of the exploration—invention—discovery learning cycle in the guide? How about the process skills? Can you find examples of how they are taught or used?
7. What does the teacher do in SCIS? Can you use this information to help you become a better teacher? Think about ways you can use the SCIS approach to develop your own science program.

C. You have had the opportunity to review some SCIS guides and develop some opinions. Write down some of your ideas or impressions. These notes should guide you in your future thoughts about SCIS.

D. If you have not done so, meet with your group and discuss your reviews. Remember, a discussion should involve all participants, not just a few. Participate, don't just listen. You may learn from the others if you listen only, but they won't learn anything from you. Make notes of your discussion.

NOTES

SELF-CHECK √

How well did you look at the guides? Skimming doesn't really help a lot. How were the guides similar? You should have noticed that the introductory information was generally the same in all of the guides and that the difference came in the "Design and Use of the Kit" and the "Activities" sections. Were you able to find activities that you might like to teach? Think about how you would use them and where you would get the materials. How would you like to teach this program? How do you think a child would react to the program?

COMMENTS

SUMMARY

You have reviewed some of the life science and physical science teacher's guides for SCIS. You read the guides and reacted to them in a small group discussion. This should reinforce what you read about SCIS in the background information. Be especially aware of the exploration—invention—discovery learning cycle presented in the guides.

These teachers are setting up the SCIS activity Life in an Aquarium.

Activity 2: Activities for SCIS

Four SCIS activities follow to give you the opportunity to participate in some of the activities you have been reading about. There are two life science activities and two physical science activities, and you should do at least one of each. Simple equipment, usually found in any elementary classroom or easily obtainable, is used in all of the activities. Each one was adapted from actual SCIS activities and uses the same type of teaching strategy as SCIS. Look for examples of exploration—invention—discovery as you do the activities.

You are to work with a small group on these activities. You can work alone, but much better results will be obtained from group interaction. Before selecting the activities your group will do, read through all four of them. Your group might want to divide, if it is large enough, and have each group do two activities so that you can share in all four of the activities. When working on the activities, don't be afraid to ask "What would happen if . . . ?" and then find out. You are not limited to the questions and procedures presented here.

Think about how a child might react to each activity. In trying to get ideas to use as a teacher, don't forget the student's viewpoint. Refer to the material you have read and the guides you have reviewed to help you put the entire program into perspective. Talk to your group and to your instructor to be sure that you understand the SCIS philosophy and what is really being taught in an activity.

Life Science Activity 1: Life in an Aquarium

The aquarium is a central part of many life science activities throughout all grade levels of SCIS. It offers an ideal, controllable environment for observation and experimentation. This activity is used to introduce first grade students to living organisms and to allow sixth grade students to experiment

Students can make many observations by studying an aquarium.

with ecosystems.[16] It is a very versatile activity, requiring only an aquarium and a little imagination.

A. An aquarium is a common fixture in many elementary classrooms. It is not expensive and can easily be set up and maintained. You have, no doubt, seen an aquarium. But have you ever really looked closely at one? You will get a chance to do so in this activity. Get an aquarium and you will be ready to start.

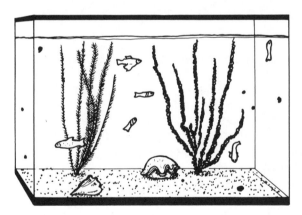

B. Observing an Aquarium

1. Obviously, the first thing you will notice in an aquarium is the animal life. Usually this will be the fish. Look closely at them.
   How many are there?   Are they all the same kind?   If not, how many kinds are there?   How do you know that they are not all the same species?

---

[16]Adapted from University of California, SCIS unit *Organisms* (Chicago: Rand McNally & Co., 1970).

2. Look for other forms of animal life. Do you find any?   What are they? What do they contribute to the system?

3. Animal life is not the only kind of life found in an aquarium. Most aquariums will also have plant life. How many forms of plant life can you find?   Describe them.

4. Are plants really necessary for a well-ordered aquarium?   What do you think the plants contribute to the system?

5. A certain amount of nonliving material can also be found in an aquarium. The most obvious, of course, is the water and the container. What other nonliving materials can you find? Describe each and tell what you think it contributes to the system.

6. The container and everything in it make up a habitat for the animal life found there. Can you describe a system that makes up *your* habitat?

C. Observing a Specific Organism

1. In section B of this activity you looked at the total system and its component parts. Now you are going to be more specific. Pick out one organism and observe it. You may want to use a hand lens to help you see it better. *Do not* remove the organism from its habitat.

2. Which organism did you pick? Describe it. (A sketch might be appropriate to aid your description).

3. How does the organism move about?

4. How do you think it eats?

What does it eat?

D. Using the Aquarium in the Classroom

1. Two problems are presented to you as a teacher. Read both of them. Then, with your group, choose one for discussion. After your group has discussed the problem, describe your course of action.

   a. What would you do if you came into your classroom and found a dead fish in your aquarium?

   Did your group perceive the dead fish as an asset or a liability? SCIS considers the event a teaching opportunity because the students can try to find out why the fish died. They can also leave the fish in the container or put it in a separate, special container and then watch decomposition.

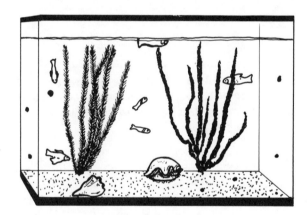

   b. How could your students find out where the "black stuff" (detritus) on the bottom of the tank comes from? The "black stuff" or detritus is waste material that comes from the organism in the aquarium. Did you think to devise experiments to isolate each organism from all of the others to see if it produced the "black stuff." This activity offers a good opportunity to devise and carry out experiments.

2. Develop a problem of your own that you might use in your classroom.

√ SELF-CHECK

For the activity:

Have you ever looked closely at an aquarium before? There is more to it than pretty fish swimming. Did you use the observation skills you worked on in Part 1-1, "Sciencing?" Could you tell the difference between the male and female organisms? Some organisms have easily recognized differences, but others do not. You might want to look at the *Organisms* guide now that you have completed the activities.

Implication for teaching:

Could you find a use for an aquarium in your classroom? An aquarium, like animals in the classroom, can either be used as a teaching tool or as a "pet" with no real teaching purpose. The choice is up to you.

COMMENTS

SUMMARY

Life in an Aquarium introduced you to the many possible uses of a common piece of classroom equipment. You were asked to make some observations and inferences to help you understand how SCIS uses a simple piece of equipment throughout the program.

Life Science Activity 2: Field Trip

This activity is a combination of several SCIS activities modified to be used as a modular, self-directed exercise.[17] In the actual SCIS program, the outside classroom is used quite often in both physical science and life science units, which are teacher directed but student oriented. As you do this activity, think about all of the ways you could use the outdoor laboratory in your school. Every school has one available, even those in the core of the city. It may be small or paved over, but it is there for your use.

A. You are going to go outside for this activity. A field trip does not have to be an elaborate venture, as you will see. You do not need much equipment for this activity but you should have a hand lens. (Every child should have access to a hand lens.) You should have a small container if you wish to collect specimens on your outing.

---

[17]Adapted from University of California, SCIS unit *Environments* (Chicago: Rand McNally & Co., 1970).

Examining organisms in the outdoor laboratory is important to the SCIS program.

B. Seasons

1. SCIS does not teach a unit on seasons as such, but children are aware of the changing seasons. Each season has its own characteristics; so each living organism adapts, in its own way, to those characteristics. Life, or evidence of life, can be found in all seasons. Consequently, even if the ground is covered with snow rather than flowers, this activity is pertinent.

2. What is the season?

3. What are the characteristics of this season?

4. Which season is your favorite? Why?

C. Field Trip

1. Go outside on a short walk with one or more members of your small group and select a small area for observation. This area need not be more than a few meters square. You might even do this activity on your way to or from class.

2. Look for organisms or signs of organisms. List some of your observations.

    Organisms   *Signs of Organisms*

3. Can you find evidence of a food chain? Describe it.

4. Look for evidence of humans having visited your area. Describe the evidence.

D. Habitat

1. A habitat is a place where a plant or animal lives. Many habitats can be found on your short walk.

2. Locate a habitat, either plant or animal, and describe it.

E. Discussion

1. Share your observations with others in your group. Discuss the evidence presented.
2. Invite your instructor to participate in your discussion.

√ SELF-CHECK

For the activity:

Did you have trouble finding a suitable site for your observation? You should not have because any site is a good one. If you had difficulty collecting evidence of organisms, then you didn't look closely enough. Such things as an insect's wing, a feather from a passing bird, or even a discarded can or cigarette butt indicate that some type of organism has passed through your site. Remember, look for simple signs.

Implications for teaching:

Try to get your students interested in the world outside of the classroom. Think of variations of this activity that you might use with your class. What would the teacher be doing while the students are involved in this type of activity?

COMMENTS

SUMMARY

You were asked to participate in an outdoor laboratory experience. SCIS uses many variations of this activity. It is inexpensive and has unlimited possibilities for the imaginative teacher.

This student is learning about electricity through direct experience.

Physical Science Activity 1: Batteries and Bulbs

SCIS uses electricity in several different activities, each one building on the previous one. Simple electrical circuits are introduced in the second grade to illustrate interaction. More complex concepts are developed, building up to

the sixth grade unit where electrical models are invented. Think about how you might use the simple act of lighting a bulb as a teaching opportunity. Flashlight batteries provide a safe, economical power source and are usually familiar to all of the students. This activity is divided into two parts to illustrate two of the three SCIS teaching strategies: exploration and discovery.[18] You will also be engaged in some invention as you discuss what you are doing within your group, but it will not be instructor guided unless you ask your instructor to join you.

A. Lighting a Bulb.

1. You will need a battery, a bulb, and two pieces of wire. Obtain these from the instructor or materials center.
2. Your first task is a simple one. You are to make the bulb light up. Use any or all of your materials. *Warning:* Stay away from wall outlets. Use only the dry cell battery for a power source.
3. Draw a picture of the way you arranged your equipment to make the bulb light up.

4. Can you devise an alternative method of lighting the bulb? *Hint:* It can be done using either one wire or two. If you used one wire in 2 above, try using two wires or vice versa. Draw a picture of your second method.

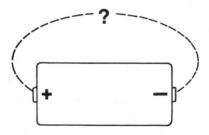

---

[18]Adapted from University of California, SCIS unit *Model: Electric and Magnetic Interactions* (Chicago: Rand McNally & Co., 1970).

5. Look for a circuit in your two drawings. This is the path that the electricity travels to get from one end of the battery to the other. If you are not sure of what a circuit is, check with your instructor. Show the circuit on your drawings.

6. Compare your results with others in your group. How did you feel when you finally got your bulb to light up? Discovery is a great feeling, isn't it?

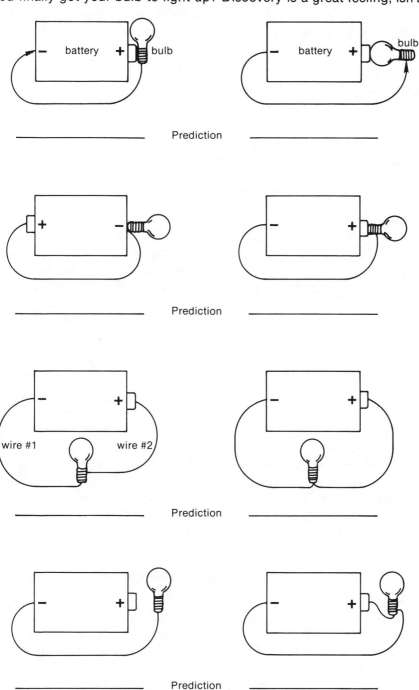

_____ Prediction _____

_____ Prediction _____

_____ Prediction _____

_____ Prediction _____

B. Solving a Problem

1. On your own, make a prediction about whether or not the bulb will light in each of the following pictures. Don't check it yet, just predict.
2. Compare your predictions with those of the other members of your group. If you disagree, can you convince them that you are right?
3. Using your materials, go back and test each circuit. Mark those that you had trouble with. Have a group discussion about this activity. Ask your instructor to participate.

√ SELF-CHECK

For the activity:

Did you have any trouble getting the bulb to light up? Wasn't it a great feeling when you finally did it? Did you check all of your predictions? It is easy to do when you have the actual equipment in your hands. You get much more positive feedback this way instead of using an abstract answer sheet.

Implications for teaching:

Can you think of other ways to use this activity? Simple materials can be used a variety of ways if you use your imagination. Think about safety as you plan activities for your students. This activity could be dangerous without safety warnings about using the proper power source, but is harmless if proper precautions are taken. Safety is too often neglected in the classroom.

COMMENTS

SUMMARY

Children use electricity all the time. You enter a room and flip a switch, and the light comes on, or turn on a flashlight, and it produces light. But why or how? Here is a common phenomenon that nearly all children are familiar with. Why not use it as a teaching device? Electricity is used in the second grade, third grade, and sixth grade in the SCIS program.

Physical Science Activity 2: Sorting Rocks

The concept that the word *object* refers to a piece of matter and that objects can be described by their properties is the first thing taught to elementary children in the SCIS physical science program. In *Sorting Rocks* you will see how a child describes the properties of rocks and learns that a single object might be made up of more than one material.[19] Language development is also

---

[19]Adapted from University of California, SCIS unit *Material Objects* (Chicago: Rand McNally & Co., 1970).

stressed in this activity because the children are encouraged to discuss their observations and conclusions.

A. Describing Properties

1. To begin this activity, each member of your group will need about five or six rocks. Obtain these from the rock box or wherever you can find five or six good rocks. Try to get a variety. A hand lens will be helpful, if available.
2. Carefully look at all of your rocks. Spend several minutes and really look closely.
3. Pick out your favorite rock and describe it to the rest of your group. Summarize your description below.

The unit Sorting Rocks gives an introduction to one area of physical science.

4. Some common properties that an elementary student might use to describe rocks are these: shape, size, color, and texture. See if you can sort your rocks using some *other* identified properties. Make a list of these properties.

5. Show your list to the others in your group, along with your collection of rocks. Can they match each rock to your description of it? If not, modify your list of properties so they can.

B. Sorting Rocks

1. To start this activity, combine your set of rocks with those of the other members of your group. This will be a group activity.
2. Sort the rocks into piles according to various properties. Each member of the group will sort the rocks, using a specific property. You are to list the different properties used by your group. *Note:* For fun, each group member might sort the rocks into two or more piles using a property known only to him. The other members then try to guess what property was used. This is not a part of the SCIS, but it is an example of a variation that you might use if your children were interested.

√   SELF-CHECK

For the activity:

In A, did you use only properties when you described your favorite rock, or did you sneak in some functional description? Did all of your rocks fit neatly into categories, or did you have trouble with some because of multiple characteristics? In B, did you have any trouble thinking of a property not used by others? You might want to look at the *Material Objects* guide.

Implications for teaching:

Where would you obtain rocks for this activity? Any roadside or stone pile has a good selection. Try to use rocks large enough to be easily handled, about two or three centimeters in diameter. Can you devise an *invention* lesson to go with this activity?

COMMENTS

SUMMARY

The first step in this physical science portion of SCIS is to learn what an object is and that it can be described by its properties. Rocks are used as objects to be described in this activity because they are easy to obtain and come in great variety.

Summary: SCIS

This is the end of the *Science Curriculum Improvement Study* portion of Part 2-1. The "Background Information" section provided you with material about the original SCIS program, supplemented with information about the two new revisions *SCIIS* and SCIS II. You also found out how these programs are taught. A review of several teacher's guides gave you an opportunity to see firsthand what information the teacher has when she teaches the program. The activities gave you hands-on experience with SCIS type exercises. Furthermore, you were asked to look for ideas that will help you develop your own teaching strategies. SCIS presents a unique learning cycle, exploration—invention—discovery, that you should be familiar with.

Now that you have completed this section of Part 2-1, you should be able to do the following:

1. Identify and describe the materials in the *Science Curriculum Improvement Study* (SCIS) program and its revisions *SCIIS* and SCIS II
2. Identify and describe the scope, sequence, and teaching strategy of the *Science Curriculum and Improvement Study* and its revisions

If you doubt that you can perform these objectives, see your instructor for help. Otherwise, you are ready for the final seminar and instructor check.

FINAL SEMINAR

Review the activities you have completed, making note of any questions that you may have. Meet with a small group of your classmates to go over the SCIS program. Be sure that you are aware of the new revisions (*SCIIS* and SCIS II) of the basic SCIS program. *Ask your instructor to participate in this final group seminar.*

NOTES

BIBLIOGRAPHY

Evaluation Supplements. Berkeley: Science Curriculum Improvement Study, University of California, 1974.

Karplus, Robert, and Thier, Herbert. *A New Look at Elementary School Science.* Chicago: Rand McNally & Co., 1967.

_____ . *SCIS Teacher's Handbook.* Berkeley: Science Curriculum Improvement Study, University of California, 1974.

Paldy, Lester et al. *SCIS II Sample Guide.* Boston: American Science & Engineering, 1978.

Science Curriculum Improvement Study. (Promotional Brochure R10/71 90281) Chicago: Rand McNally & Co., 1971.

SCIS Final Report. Berkeley: SCIS, University of California, 1976.

SCIS Teacher's Guides. Chicago: Rand McNally & Co., 1970.

Thier, Herbert et al. *Teacher's Guides for SCIIS.* Chicago: Rand McNally & Co., 1978.

SUGGESTED READING

Carin, Arthur A., and Sund, Robert B. *Teaching Modern Science.* 2d ed. Columbus, Ohio: Charles E. Merrill Publishing Co., 1975. Pp. 51-55.

Gega, Peter C. *Science in Elementary Education.* 2d ed. New York: John Wiley & Sons, 1970. Pp. 536-47.

Karplus, Robert, and Thier, Herbert. *A New Look at Elementary School Science.* Chicago: Rand McNally & Co., 1967.

Renner, John W., and Ragan, William B. *Teaching Science in the Elementary School.* New York: Harper & Row, Publishers, 1968. Pp. 258-79.

SCIS Final Report. Berkeley: SCIS, Univeristy of California, 1976.

Science . . . A Process Approach II

Background Information

Author: Developed by the Commission on Science Education of the American
Association for the Advancement of Science (AAAS)
1515 Massachusetts Avenue, N.W.
Washington, D.C.

Publisher: Ginn and Company
A Xerox Educational Company
191 Spring Street
Lexington, Massachusetts 92171

History

In 1961, a group of science teachers and scientists working with staff members of the Commission on Science Education of the American Association for the Advancement of Science (AAAS) undertook the task of developing a new elementary science program. The National Science Foundation encouraged the endeavor by providing necessary financial support. The format of the new program was the focal point of early discussions. Several proposals such as "major themes,"or "selected topics for each grade," or "concrete and familiar" were suggested, but the decision was made to go with the teaching of the scientific processes. The name of the new program, *Science . . . A Process Approach* (SAPA), reflects this decision.

After the preliminary discussions and other basic decisions regarding the format, the hard work started. Writing teams were established, and they held four summer workshops to write the program. Trial centers were set up across the nation for field

This preservice teacher is preparing an SAPA II lesson.

testing. After six years of writing, testing, and revision, the SAPA program was ready for public use. As the program was used and time passed, the need for a revision became evident. Important changes in science education, such as emphasis on teaching flexibility and the need for individualized learning, contributed to the need for revision as did the development of a deeper national concern for the environment. The AAAS Commission on Science Education again accepted the challenge and undertook the task of revising the program. This time, Ginn and Company provided the financial support.

Input for the revision came from many areas. The major contributors were those classroom teachers who had used the program. Their experiences were the source of invaluable insight and suggestions. Trial centers also provided information and suggestions based on their experiences with the material. The Eastern Regional Institute for Education made several studies that helped provide concrete data. Finally, several special conferences and symposia were held to obtain additional ideas and suggestions. When the consulting, rewriting, and trial testing was completed in 1975, SAPA II was ready for use.

Conceptual Background

The major factor in establishing the conceptual framework of the SAPA II program was the decision to use science processes as the basis of the program. The traditional approach to teaching science is to use subject-matter content as the base, bringing in the process skills as they are needed. SAPA II is unique in that it reverses this procedure. The process skills form the base of the program, and content is introduced as needed. The assumption is that if elementary children learn the processes, they can understand the content.

All together, thirteen processes were identified to be used in the program. In the early grades, K-3, the following eight basic processes are developed:

1. *Observing*—Children learn to use their five senses to determine properties of an object.
2. *Using Space/Time Relationships*—Children learn that five categories: shapes, direction and spatial arrangement, motion and speed, symmetry, and rate of change are used to describe spatial relationships and their change with time.
3. *Classifying*—Children learn to use classification schemes to impose order on collections of objects or events.
4. *Using Numbers*—Children learn that numbers are basic to science. Numbers allow scientists to make measurements, classify objects, and order objects.
5. *Measuring*—Children learn that they can quantify their observations by using the proper measuring device. Metric measurement is used exclusively.
6. *Communicating*— Children learn that they can use graphs, diagrams, maps, and mathematical equations as well as oral or written words to communicate.
7. *Predicting*—Children learn to make predictions based on observations, measurements, and certain inferences. They also learn the difference between a valid prediction and a guess.
8. *Inferring*—Children learn that an inference is an explanation of an observation and that more than one inference may be made in order to explain an observation.

After the child has learned to use the basic processes in the early grades, he is introduced to the integrated processes in the fourth, fifth, and sixth grades. There are five processes in this group:

1. *Controlling Variables*—Children learn to identify variables and then control them to manage the conditions of an experiment.
2. *Interpreting Data*—Children learn to use data collected to make predictions, inferences, and hypotheses. They also learn to collect data.
3. *Formulating Hypotheses*—Children learn to use observations and inferences to formulate generalizations (hypotheses) about all objects or events of the same class. They also learn to test their hypotheses.
4. *Defining Operationally*—Children learn to define terms in the context of their own experiences and to work with a definition rather than memorize it.
5. *Experimenting*—Children learn to use all of the previous processes to formulate and answer a problem.

The program developers felt that as these process skills were mastered, the student would develop intellectual skills that could be used in all areas of science. What better way for a student to learn science than to do what a scientist does?

Underlying the development of the program were five criteria considered to be basic to a quality science program. Every effort was made to make the new program meet these criteria. From the SAPA II *Program Guide*, here are the criteria:

1. Science in the elementary school must contribute to the general education of every child.
2. A viable science program must enable children to reach a high level of achievement.
3. Children must be challenged by asking them frequently to think, to reason, and to invent.
4. Children's experiences must be broadened into many fields of science.
5. A good instructional program in science must be realistic in developing intellectual skills.[20]

A third factor also contributes to the SAPA II program. The developers felt that all children should be able to experience success and that major goals could be reached if pursued one step at a time. This thinking led to the establishment of a sequential program based on several small steps, each with objectives that could be easily achieved. The goal is for 90 percent of the students to achieve 90 percent of the objectives.

What about the graduate of the SAPA II program? The end product is the real proof of the program. What should the student be able to do when he has completed all of the modules? The goals of the program state that the graduate should be able to do the following.

(a) Apply a scientific approach to a wide range of problems, including social ones, distinguishing facts from conjectures and inferences, and identifying the procedures necessary for verification of hypotheses and suggested solutions.

---

[20]From *Program Guide* of SCIENCE . . . A PROCESS APPROACH II, © Copyright, 1975, American Association for the Advancement of Science. Used by permission of the publisher, Ginn and Company (Xerox Corporation).

(b) Acquire an understanding of the sciences he or she will pursue in junior and senior high school more rapidly and with less difficulty than would otherwise be possible.

(c) Identify each of the following in a printed or oral account of a scientific experiment: the question being investigated; the variables manipulated, controlled, and measured; the hypothesis being tested; the relationship of such a test to the results obtained; the conclusions that can legitimately be drawn.

(d) Infer, where necessary, the question being investigated and the elements of scientific procedure in an incomplete account of a scientific experiment, such as might appear in a newspaper.

(e) Design and, under certain conditions, carry out one or more experiments to test hypotheses relevant to a problem, providing the problem is amenable to scientific investigation and the content is within the child's understanding.

(f) Show appreciation of, and interest in, scientific activities by choices made in reading, entertainment, and other kinds of leisure-time pursuits.

(g) Develop values, attitudes, value systems, and value judgment criteria not only applicable to science-related experiences but transferable to day-to-day experiences throughout life.[21]

In summation, the conceptual framework of the SAPA II program can be described as a conscientious effort to develop a quality science program based on the systematic teaching of the process skills.

Program Description

SAPA II is unique in that it is a sequential, but modular, program based on the assumption that all children, no matter where they may live, need to learn the processes of science. If school personnel choose to use this program, they are committed to the process philosophy for all grade levels up through the sixth grade. It cannot be used piecemeal because each activity builds on previous activities. However, there is an exception. If teachers cannot, or do not, want to change their science program but do want to include some process development, they can use some of the SAPA II modules to supplement their program. A module can be described as a unit of work involving a specific phase of a specific scientific process. In SAPA II, the unit of work consists of several, usually three or four, activities. The specific process taught may be any of the eight basic skills or five integrated processes. The specific phase refers to the concept of logical progression. Each process is taught in several steps, each step building on the step previously presented.

Each module is designed to achieve specific objectives. There may be as few as one or as many as five, but every objective is important in the overall sequence of the program. To keep the objectives simple and easily attainable, SAPA II uses 15 modules for each grade level, or a total of 105 modules. A list of these modules follows.[22] As you look at this list, keep the following in mind:

1. Each module teaches only one process.
2. There are fifteen modules per grade level.

---

[21]From *Program Guide* of SCIENCE . . . A PROCESS APPROACH II, © Copyright, 1975, American Association for the Advancement of Science. Used by permission of the publisher, Ginn and Company (Xerox Corporation).

[22]Adapted from *Curriculum Catalog* of SCIENCE . . . A PROCESS APPROACH II,© Copyright, 1976, American Association for the Advancement of Science. Used by permission of the publisher, Ginn and Company (Xerox Corporation).

3. The subletters beside each process denotes the number of that process; e.g., /c means that this is the third module of the particular process.
4. Modules in each stage can be taught in any order, but each stage must be completed before going on to modules on the next stage.
5. Ideally, modules should be taught in numerical sequence.
6. Asterisked numbers indicate the better modules for individualized usage.

Modules for SAPA II

Grade Level	Stage	Module No.	Process	Module Title
K	I	1	Observing/a	Perception of Color
		*2	Space/Time/a	Recognizing and Using Shapes
		*6	Space/Time/b	Direction and Movement
		7	Observing/d	Perception of Taste
	II	3	Observing/b	Color, Shape, Texture, & Size
		5	Observing/c	Temperature
		8	Measuring/a	Length
	III	4	Classifying/a	Leaves, Nuts, & Seashells
		9	Using Numbers/a	Sets and Their Members
		*12	Space/Time/d	Three-Dimensional Shapes
		*11	Observing/e	Listening to Whales
	IV	10	Space/Time/c	Spacing Arrangements
		13	Using Numbers/b	Numerals, Order, & Counting
		14	Classifying/b	Animal & Familiar Things
		15	Observing/f	Perception of Odors
1	V	*16	Classifying/c	Living & Nonliving Things
		17	Observing/g	Change
	VI	18	Observing/h	Using the Senses
		19	Observing/i	Soils
		20	Using Numbers/c	Counting Birds
		23	Measuring/b	Comparing Volumes
		25	Communicating/b	Introduction to Graphing
	VII	*21	Observing/j	Weather
		22	Communicating/a	Same but Different
		*24	Measuring/c	Metric Lengths
		*26	Measuring/d	Using a Balance
	VIII	27	Communicating/c	Pushes and Pulls
		28	Observing/k	Molds & Green Plants
		29	Space/Time/e	Shadows
		30	Using Numbers/d	Addition Through 99

Grade Level	Stage	Module No.	Process	Module Title
2	IX	31	Communicating/d	Life Cycles
		*32	Classifying/d	A Terrarium
		33	Inferring/a	What's Inside
		*34	Measuring/e	About How Far?
	X	36	Observing/l	Animal Responses
		37	Measuring/f	Forces
		39	Measuring/g	Solids, Liquids, and Gases
		*40	Inferring/b	How Certain Can You Be?
	XI	35	Space/Time/f	Symmetry
		38	Predicting/a	Using Graphs
		42	Classifying/e	Sorting Mixtures
		43	Communicating/e	A Plant Part That Grows
	XII	41	Measuring/h	Temperature & Thermometers
		44	Predicting/b	Surveying Opinion
		*45	Space/Time/g	Lines, Curves, & Surfaces
3	XIII	*46	Inferring/c	Observations & Inferences
		47a	Communicating/f	Scale Drawings
		*48	Predicting/c	The Bouncing Ball
		49	Measuring/i	Drop by Drop
	XIV	47b	Communicating/g	A Tree Diary
		*50	Predicting/d	The Clean-Up Campaign
		54	Measuring/j	Static & Moving Objects
	XV	52	Inferring/d	Plants Transpiring
		53	Predicting/e	The Suffocating Candle
		55a	Observing/m	Sprouting Seeds
		55b	Observing/n	Magnetic Poles
	XVI	51	Space/Time/h	Rate of Change
		56	Classifying/f	Punch Cards
		58	Inferring/e	Liquids & Tissue
	XVII	57	Communicating/h	Position & Shape
		*59	Using Numbers/e	Metersticks, Money, & Decimals
		60	Space/Time/i	Relative Motion
4	XVIII	*61	Inferring/f	Circuit Boards
		62	Controlling Variables/a	Climbing Liquids
		63	Interpreting Data/a	Maze Behavior
		65	Interpreting Data/b	Minerals in Rocks
	XIX	64	Defining Operationally/a	Cells, Lamps, Switches
		*66	Controlling Variables/b	Learning & Forgetting
		67	Interpreting Data/c	Identifying Materials
		70	Formulating Hypotheses/a	Conductors & Nonconductors

Grade Level	Stage	Module No.	Process	Module Title
	XX	*68	Interpreting/Data/d	Field of Vision
		69	Defining Operationally/b	Magnification
		*71	Controlling Variables/c	Soap & Seeds
		*74	Defining Operationally/c	Biotic Communities
	XXI	72	Controlling Variables/d	Heat Rate
		73	Formulating Hypotheses/b	Solutions
		75	Interpreting Data/e	Decimals, Graphs, & Pendulums
5	XXII	76	Interpreting Data/f	Limited Earth
		77	Controlling Variables/e	Chemical Reactions
		78	Formulating Hypotheses/c	Levers
	XXIII	80	Defining Operationally/d	Inertia & Mass
		*83	Formulating Hypotheses/e	Angles
	XXIV	*79	Formulating Hypotheses/d	Animal Behavior
		81	Defining Operationally/e	Analysis of Mixtures
		82	Controlling Variables/f	Force & Acceleration
		*85	Interpreting Data/h	Contour Maps
	XXV	84	Interpreting Data/g	Angles
		89	Defining Operationally/g	Plant Parts
		90	Interpreting Data/k	Streams & Slopes
	XXVI	86	Interpreting Data/i	Earth's Magnetism
		87	Interpreting Data/j	Wheel Speeds
		*88	Defining Operationally/f	Environmental Protection
6	XXVII	*91	Defining Operationally/h	Flowers
		92	Formulating Hypotheses/f	Three Gases
		*95	Interpreting Data/l	Mars Photos
	XXVIII	93	Defining Operationally/i	Temperature & Heat
		94	Controlling Variables/g	Small Water Animals
		96	Experimenting/a	Pressure & Volume
	XXIX	97	Experimenting/b	Optical Illusions
		*98	Experimenting/c	Eye Power
		99	Experimenting/d	Fermentation
	XXX	100	Experimenting/e	Plant Nutrition
	XXXI	*101	Experimenting/f	Mental Blocks
		102	Experimenting/g	Plants in Light
	XXXII	103	Experimenting/h	Density
		104	Experimenting/i	Viscosity
	XXXIII	105	Experimenting/l	Membranes

In SAPA II, there is a kit for each module. Each kit contains most of the materials needed to teach the module and an instruction booklet. Such items as living organisms and perishable materials are not included. The teacher must arrange to purchase them locally as well as the more common items usually found around the classroom, such as construction paper, rubber bands, and scissors. On the other hand, most of the necessary laboratory materials, such as containers, metersticks, metric rulers, washers, chemicals, and thermometers, are included. Printed material furnished in the kit includes pictures, charts, transparencies, and thirty nonconsumable student booklets. Spirit duplicator masters are also furnished so that the teacher can reproduce necessary worksheets. Some items are consumable and must be replaced as they are used. If kits are shared, each teacher will need extra packages of the consumables for the classroom. These are available for reorder.

The contents of a SAPA II kit vary with the topic. Some modules are supplied with many materials; whereas, other modules do not need much. For example, module 46, *Observations and Inferences,* needs only a small amount of equipment. It has thirty cartoon booklets, three wall charts, ten footprint cutouts, thirty worksheets, and five nutcrackers. On the other hand, a module like module 103, *Experimenting,* needs more equipment: modeling clay; equal arm balances; five sets of gram masses; graduated cylinders; iron, lead, and aluminum samples; marbles; graph paper, and two kinds of booklets. The best way to know exactly what a kit provides is to look at some of the instruction booklets. You will have the opportunity to do this when you start the SAPA II activities.

The module instruction booklet is a unique feature of the program. It is a guide that tells the teacher what the child is expected to learn, why it is important, how to teach it, what materials are needed, and how to find out if the child learned what was taught. The teacher follows the booklet closely, or uses it as a guide, depending upon her experience and the needs of the students. Theoretically, if the teacher follows the instructions in the booklet, 90 percent of the students will achieve 90 percent of the stated objectives for the module. Each instruction booklet contains the following information:

1. *Objectives*—those minimal behaviors expected of each student who completes the module
2. *Sequence Chart*—shows the prerequisites for the module as well as what comes next (A large planning chart is available to show where each segment fits into the overall program.)
3. *Rationale*—background information and sometimes advice on the module
4. *Vocabulary*—new words that will be used in the module
5. *Instructional Procedure*—the actual teaching strategy (Included are an introduction, several activities, and a material list for each activity.)
6. *Generalizing Experiences*—activities designed to allow the student to relate what was learned to different situations
7. *Appraisal* (Modules 1-60)—a group activity used to determine if the students have met the objectives
8. *Group Competency Measure* (Modules 61-105)—a group activity used to determine if the students have met the objectives
9. *Competency Measure*—individual testing procedure using test items directly related to each objective

The cost of the SAPA II program is comparable to other science programs. You can order the program as individual modules, by grade level, or as a total system. The cheapest way, of course, is to implement the total program. Remember that there is a yearly maintenance cost for expendable items as well as additional costs for living materials and perishable items. An actual cost breakdown can be found in Ginn and Company's promotional catalog. Check with your instructor to see if one is available.

Teaching Strategy

The SAPA II teacher acts as a guide, instructor, evaluator, and resource person. As a guide, it is the teacher's task to keep the students moving toward the goals of the program. She may allow them to wander a little, exploring side topics, but keeps them moving toward their goals. This is done by asking questions, pointing out new information, presenting new material, or guiding discussions. As the instructor, she must also act as an evaluator. In this role, she keeps a progress record for each child and determines what needs to be done next. The teacher, as a resource person, answers student's questions or sends them to appropriate sources, such as charts or books, or helps them set up an experiment.

The students also have a role in SAPA II. They are learners and, as such, are actively involved in the learning process. They must learn the process skills for themselves, with the teacher's help. Consequently, they do activities, discuss their ideas, and question what they have done. Faster students participate in optional activities, extending their understanding. Students also are involved in peer teaching, helping each other when necessary. They get their hands on the program, use the teacher when necessary, and learn. Obviously, students must be motivated to want to assume this role. SAPA II thinks that involvement in, and success with, the program provide the challenge and rewards needed to provide this motivation.

How is a SAPA II module taught? The first step, of course, is for the teacher to read over the material in the instruction booklet, becoming familiar with the objectives, the activities, and the appraisal. The instruction booklet gives explicit instructions as to what to do and how to proceed. If necessary, he might also go through the activities to see if there are any problems.

To teach a module, the teacher should follow the directions in the instruction booklet as to how to introduce the unit and how to proceed through the activities. The booklet gives explicit directions: "Do this, then ask these questions," or "Have the students do that, then ask for their responses." Look for this when you review the SAPA II booklets in the "Activities" section. The teacher can vary the approach or the questions asked if a change seems warranted.

After teaching the module, the teacher evaluates the students. Two evaluation activities are given in the instruction booklet—an appraisal and a competency measure—with specific directions. The results are recorded, and the module is completed. Teacher and students are ready to move on to the next module.

Evaluation of Student Performance

The evaluation of student performance is very structured in SAPA II. Specific behavioral objectives are given at the beginning of each module and are tested for at the end. If the student can perform them, he has achieved; if not, he hasn't achieved. Actually, this is an oversimplification of the suggested method of evaluating student performance, but it is the central idea.

The teacher must do several things to evaluate a student's performance. He has to make ancedotal notes during the lessons and administer the "Appraisal Activities" and/or "Group Competency" measures. Individual records must also be kept for each student. This is usually done on SAPA II "Tracking Cards." These cards provide a continuous record of each student's performance from the time she first enters the program until she completes it.

Specific recommendations for evaluating student performance are given by SAPA II in the *Program Guide:*

1. Use the *Sequence* chart on the first page of each *Instruction Booklet* to determine whether or not the children have the prerequisite skills; if not, provide the appropriate DEBUT modules, or review the module objectives identified as prerequisite.

2. After the modules have been taught, administer the *Appraisal* activity (Modules 1-60) or the *Group Competency Measure* (Modules 61-105) to measure the achievement of objectives by a group of children.

3. Administer the *Individual Competency Measure* to a small group of pupils for each module. In selecting students, you may wish to include some from the high- and some from the low-achiever groups and some from the group that you feel least certain about. This evaluation device may be waived if evidence warrants it.

4. Keep an individual profile sheet for each pupil, even though you do not have a competency score on each objective for each pupil. Many tasks can be scored from your observation of investigations by individuals and small groups of children, without use of the *Competency Measures. Tracking Cards* can be used here.

5. Keep an anecdotal record or diary of significant events in your teaching (failures as well as successes), including the involvement of individuals by name.

6. Administer check-point tests and end-of-the-year science tests if you like. Choose tests that measure individual performance rather than memory of facts.

7. Report to parents on the progress of their children in achieving competence in the processes of science. *Tracking Cards* may be discussed in conferences with parents. In the periodic reports of student progress made in most schools, inclusion of statements about skills in the processes of science would be interesting and helpful to parents.

8. Encourage parents to report their children's comments on science experiences in school.

9. If your school requires reporting by letter or numerical grade, the *Tracking Cards* will provide an adequate basis for determining the children's grades. Devise scales that are appropriate for your school and for your groups of children.

10. Finally, the anecdotal record referred to in Step 5 can serve as a basis for describing student interests, attitudes, and character development. The use of varied instructional modes in SCIENCE . . . A PROCESS APPROACH II will help children see that learning can be fun, and that it sometimes proceeds best when you work with friends in one way or another.[23]

---

[23]From *Program Guide* of SCIENCE . . . A PROCESS APPROACH II, © Copyright, 1975, American Association for the Advancement of Science. Used by permission of the publisher, Ginn and Company (Xerox Corporation).

Implementation

Implementation of the SAPA II program is not at all complex. The first stage, as mentioned earlier in the ESS and SCIS programs, is to get the teacher ready for the program. Inservice training is recognized as an important part of the SAPA II program, and several provisions are made for this by Ginn and Company.

It is suggested that the SAPA II program be implemented totally at one time. This method allows all students to be involved in the program at once and causes fewer problems in the long run.

Teachers will have to provide extra background information to help the students develop the skills needed to master new modules. But, by using this implementation approach, everyone will be in sequence at the beginning of the second year.

Summary

The *Science . . . A Process Approach II* (SAPA II) program is a sequential, but modular, science program developed by the Commission on Science Education of the American Association for the Advancement of Science (AAAS) and distributed by Ginn and Company. It is based on the assumption that young children should be taught how to *do* science rather than science subject-matter content. Eight basic processes are taught from kindergarten through the third grade, followed by five integrated processes in the fourth, fifth, and sixth grades. Children are exposed to subject matter, but knowledge of the processes is the prime objective. Fifteen modules per grade level are taught in sequence with slight variation permitted. The teacher presents the material, leads discussions, evaluates the students, and keeps records of student achievement. The kit comes packaged in individual modules with materials for thirty students and can be shared by several teachers, providing each has a set of consumable materials. A module instruction booklet provides behavioral objectives, activities, and appraisal for each module. Teacher training material is available for purchase from Ginn and Company. The program has been proven effective, generally reaching its goal of 90 percent attainment of the written objectives by 90 percent of the students, if taught as suggested.

Have you read the background information for SAPA II? If not, go back and do so. Then you will be ready to do the activities. You are now going to get the opportunity to handle some of the materials you have read about.

As you proceed through the activities, you will be asked to do three things. First, you will review some of the instruction booklets. This will give you a chance to actually see the booklets and compare them to what you have read about them. Refer to the background information at any time. Second, you will do some sample activities to see what kinds of topics and activities are used in SAPA II. This will also help you to see how the processes are taught. And third, you are to think about how you could use some of the activities or ideas in your own classroom.

Look for several things as you go through the activities. For instance, how is the program structured? What material is needed to teach an activity like this? What does the teacher do? Don't forget the questions you may have had when you were reading the background material. See if you can answer some of them.

Remember, it is important to learn about SAPA II not because it may be in use in the school where you will teach, but because it offers a teaching strategy, organization, and ideas that you can incorporate into your teaching.

Activity 1: Review of SAPA II Instruction Booklets

A. In this activity, you are going to review some of the SAPA II instruction booklets. The purpose of this review is to do the following:

1. Familiarize you with the instruction booklets of SAPA II by having you examine the actual materials
2. Help you better understand how the program is structured by having you look at several booklets to see how one builds on the other
3. Help you see how the material is taught by reading through several of the activities in a booklet
4. Give you some ideas that you might use when you teach by showing you how a program can be organized and by exposing you to a source of activities that you might consider when you start teaching

Try to keep these purposes in mind as you go through the instruction booklets. Now you are ready to begin.

Obtain several SAPA II instruction booklets for review. Be selective in your choices. You might want to review several booklets that describe the same process to see how the process is developed. Other members of your group might select booklets that teach different processes. Plan to review the booklets by yourself and then discuss your observations with your group and your instructor.

B. Start at the front of the booklet and locate the module number, title, and the process being taught. You probably have already done this when making

Various SAPA II instruction booklets.

your selection. Then, read through the booklet, looking especially for the following information:

1. Compare the objectives with the activities and the appraisal. Can you see how the teacher knows what she is trying to teach, how she is going to teach it, and how she will find out if it has been taught? This is an important sequence in teaching any lesson. Can you use this information when you design a lesson?
2. Find the rationale. Does this help you understand why a particular lesson is being taught? Do you need a rationale when you teach a lesson?
3. Read through the activities. Could you use some of these ideas in your classroom if you did not have SAPA II? What kinds of materials would you need? How hard would it be to obtain the necessary materials?
4. Try to determine the teacher's role in teaching this material. What does the teacher do when she teaches? Can you teach that way?
5. Look for any ideas, methods, or activities that you might use when you start teaching. You may not have the program available to you, but you can always adapt some of the teaching strategies or activities you learn. Look beyond what the program is; look to how a knowledge of the program can improve your teaching ability.

C. Now, make some comments on your review for future reference. These comments are for your use, not for your instructor's.

COMMENTS

D. Have a group discussion on the instruction booklets. You might want to invite your instructor to participate. Talk about what you like and don't like, what you can use and can't use, and whether or not you would like to teach using SAPA II. Make some notes on the group discussion.

NOTES

√ SELF-CHECK

Did you really look at the booklets, or did you just skim through? Remember, you will get out of this activity only what you put into it. Did you compare several booklets? Even though the format is the same in all of the booklets, there are many different ideas and teaching strategies used in SAPA II. Did you refer to the background information to see what was said about the booklets? It might help make both the readings and the review more meaningful. Most important of all, did you find something to take with you when you teach? If not, look again. Even if you don't particularly like the program, there are still ideas that you can use when you develop your own program.

COMMENTS

SUMMARY

The instruction booklets tell you how to teach SAPA II. Your review should give you a better understanding of the program and some new ideas about organizing and teaching a lesson. Sharing thoughts with your group and instructor should help you get a better perspective of SAPA II.

Activity 2: Activities for SAPA II

You are now ready to participate in some SAPA II activities. There are four activities presented, and you should do at least two of them. These particular activities were selected because they use simple, readily available material. Each one involves a different process and represents a different grade level. These are actual SAPA II activities, but they have been modified slightly to make them appropriate for your use.

Read over all of the activities before selecting the ones you are going to do. Discuss the choices within your group and decide which ones you are going to do. Your group may choose to do several with all members participating in each activity, or it may choose to split into two groups, each

SAPA II activities emphasize the importance of student interaction.

doing two activities then comparing the results. For these activities, it is better to work in small groups than to work alone. Interaction among students is important in SAPA II and in these activities.

These activities will teach you how a student sees the program. You will use the same materials the student uses and answer some of the same questions. You have looked at SAPA II as a teacher, but not as a student. Try to think how, for example, a fourth grader might respond to such activities.

As you go through these activities, think about what you have read in the background information and in the instruction booklets. How does it all fit together? What do you really know about SAPA II, and how can this knowledge help you to be a better teacher?

Module 29: Using Space/Time Relationships/e
Shadows

This activity is an adaptation of a first grade activity.[24] The children have already been exposed to three-dimensional shapes and objects in previous modules. They learned to identify shapes by touch and/or sight. Now they will begin to extend this ability by learning to recognize the relationship between two- and three-dimensional geometric shapes. Shadows are used to show two-dimensional shapes of three-dimensional objects.

A. A shadow box is a simple device consisting of two basic components: a light source and a translucent screen. A filmstrip or slide projector is an ideal light source. Set it up about 1.5 meters from your screen for a clean, sharp shadow. A translucent screen can easily be constructed. All you need is a sheet of typing paper and cardboard frame to hold the paper steady. Place the object to be viewed between the light source and the screen; then view the shadow from the opposite side of the screen. You can draw on the screen,

[24]Adapted from *Module 29: Using Space/Time Relationships/e* of SCIENCE . . . A PROCESS APPROACH II, © Copyright, 1974, American Association for the Advancement of Science. Used by permission of the publisher, Ginn and Company (Xerox Corporation).

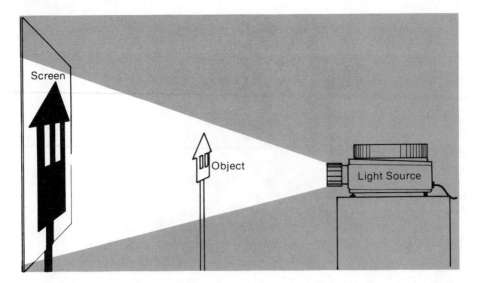

replacing it as you change objects. An overhead projector can also be used effectively as a shadow box. Simply place the object on the projection surface and turn on the light. The shadow will be projected. You might project a small image and have the children draw around the shadow. Your imagination is the only limiting factor in using the shadow box or overhead projector.

B. Have one member of your group operate the shadow box or overhead projector. The rest of the group should look at the shadows only, not at the object. You will need several simple three-dimensional objects. The operator does the first part of the directions, and the rest of the group records what they find out. The directions for this activity are as follows:

1. *Operator:* Place one of the objects in position and project the shadow.
   Group: Sketch the shadow. What do you think the object is?

2. *Operator:* Rotate the object to show another dimension.
   Group: Sketch the new shadow. Have you changed your mind about the identity of the object?

3. *Operator and Group:* Repeat the activity several times using different objects.

4. Do some objects have the same shadow no matter which way they are projected? Give an example.

5. Are the shadows of some objects similar if projected from one side but different if projected from another? Give examples.

C. You have looked at shadows and tried to picture the objects, now let's try a more abstract exercise. Here are some pictures of shadows. Try to decide what three dimensional shape was used to create the pair of shadows presented.

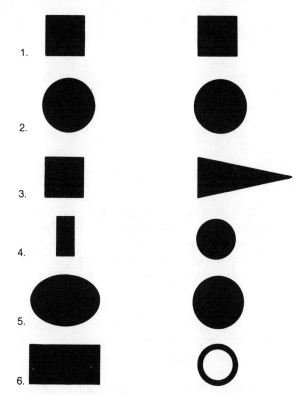

7. One of the above sets of shadows is incomplete for an accurate picture. Which one is it?   Why?

√ SELF-CHECK

For the activity:

Did you peek at the object when you were looking at the shadows? Was the shadow what you expected? If you are unsure of your ability to visualize, try several objects on the overhead projector or shadow box. Look for objects that might reproduce the shadows in C. Is number 3 shadow a pyramid or a wedge? If the instruction booklet for this module is available, you might want to refer to it.

Implications for teaching:

Can you think of variations of this activity that you could use? Could you adapt this activity to another grade level? Remember, a good idea is adaptable to many uses and situations. What materials would you need to teach a lesson such as this? How hard would it be for you to get them together? Think about the shadow box. It is a versatile piece of equipment that has many uses in subject areas other than science. Remember it when you start teaching.

COMMENTS

SUMMARY

This first grade exercise will help the students learn how to visualize an object from its outline or shape. Visualization skills are often neglected but should not be because they are an important step in developing the ability to do abstract thinking. The shadow box used in this exercise is an excellent classroom teaching device that is both easy to make and has unlimited potential.

Module 34: Measuring/e
About How Far?

In previous modules on measuring, the students have learned how to make measurements by comparing an object directly to an arbitrary unit or standard. In this module, they find out that it is not always possible, or even desirable, to make these direct measurements. They are now going to learn how to estimate linear measurements using mental images of standard units, and so are you.[25] You will need a meterstick or similar measuring device. All SAPA II measurements are in metric units.

[25]Adapted from *Module 34: Measuring/e* of SCIENCE . . . A PROCESS APPROACH II, © Copyright, 1974, American Association for the Advancement of Science. Used by permission of the publisher, Ginn and Company (Xerox Corporation).

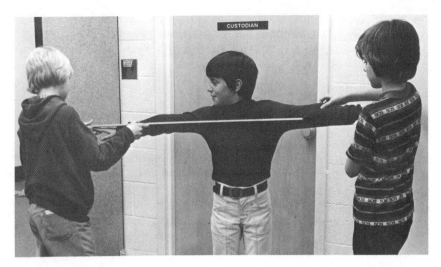

Using the length of his outstretched arms as a standard, this student will learn to estimate other linear measures.

A. Metric Reference

In order to estimate linear measurements, you need to have a few aids to help you. Using your meterstick, make the following measurements. You will have to decide upon the proper units of measurement to use. Just for fun, you might estimate these before you actually measure them.

1. Width of your finger_____

2. Length of your finger _____

3. Length of your hand _____

4. Length of your foot _____

5. Your height_____

6. Length of your pen or pencil ____

7. Length of your pace or step _____

8. Length of your arm _____

9. Length of your arms outstretched
   _____

10. _____
    (your choice)

B. Try some more metric estimations. No fair using the meterstick yet. Do only the first column.

	Estimated	Actual
1. How high is your desk?	_____	_____
2. How long is the room?	_____	_____
3. How wide is your textbook?	_____	_____
4. What is the diameter of a quarter?	_____	_____
5. How tall is your instructor?	_____	_____

Now try a couple of your own choice.

6. _____ _____ _____

7. _____ _____ _____

C. Go back and actually measure the distances that you have estimated. Compare your answers. How close were you in your estimations? If you were off by quite a bit, try a few more. SAPA II gives a general rule of plus or minus 20 percent as acceptable at this level.

√ SELF-CHECK

For the activity:

Did you have any trouble using metric measurement? Did you use your reference measurements to help you estimate distances? How many hands long is the width of the desk? How many paces long is the length of the room? What references did you use to estimate the height of your instructor? Were you within the plus or minus 20 percent allowance? If the instruction booklet for this module is available, you might want to look at it.

Implications for teaching:

How could you use this activity in your classroom? You will be teaching metric measurement; so you should start collecting metric activities as well as becoming more familiar with the metric system.

COMMENTS

SUMMARY

The student learns to estimate distances that cannot be measured directly. Mental images of common units help provide a reference base for estimations. The student is also learning to think in metric units rather than in the traditional feet and inches units.

Module 71: Controlling Variables/c
Seeds and Soap

The students learn how to control variables in this module. In modules a and b, they were introduced to the terms *manipulated variable, responding variable,* and *variables held constant.* In this module, an experiment is described by the teacher and discussed. The students must decide what the variables are, what kind each one is, and how they can modify the experiment to make it valid. You will get the same opportunity in this activity.[26] The students will have material to manipulate, but you will do your manipulating mentally.

---

[26]Adapted from *Module 71: Controlling Variables/c* of SCIENCE . . . A PROCESS APPROACH II, © Copyright, 1975, American Association for the Advancement of Science. Used by permission of the publisher, Ginn and Company (Xerox Corporation).

*The Seeds and Soap program deals with
controlling the variables in
an experiment.*

A. You need some background information before you start. *Variables* are those factors that influence an experiment. All but two must be held constant. One of two is manipulated *(manipulated variable)* and the other responds to the manipulation *(responding variable)*. Try an example. You should be aware of the well-known experiment that demonstrates a plant needs light to grow. If you were to do this experiment, you would need to control variables such as these:

1. Size and variety of plant
2. Amount of water
3. Size of containers
4. Temperature

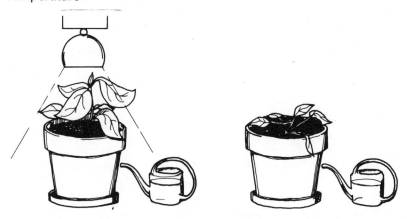

Now you are ready to *manipulate* a variable—how much sunlight each plant is to receive. One will be placed in the sun, the other in darkness. One plant grows; the other plant dies, or at least does not grow as strong and healthy as the first. Plant growth becomes the *responding variable*. Do you feel confident in your knowledge of variables? Go on to section B and test your knowledge.

B. I am a curious person who enjoys a good experiment. The other day, I accidentally poured soapy water on one of my houseplants. I wondered if I had hurt it or helped it, so I decided to set up an experiment to find out if detergent had any effect on the growth of seeds. Below is a picture of my experimental apparatus.

1.	2.	3.
5 mm vermiculite	30 mm vermiculite	20 mm shredded paper
10 mm mung bean seeds	25 mm mung bean seed	50 radish seed
10 mm vermiculite	5 mm vermiculite	10 ml water
10 ml water	20 ml liquid detergent	10 ml liquid detergent
	(Ivory)	(Joy)

Now, I will watch the containers for several days to see what will happen. The teacher would allow his students to decide whether or not this would answer the original question. Obviously it won't. Try to answer some questions about the variables in this experiment.

1. Are there *any* variables which have been held constant? Which ones?

2. List the variables in this experiment. There are fourteen.

3. Which is the manipulated variable in this experiment?

4. Which is the responding variable?

5. Do you think that my experiment is any good? Why or why not?

C. Knowing that an experiment is not valid because the variables are not controlled is not enough. You have to be able to control the variables. Go back to your list of variables and choose five. Now, list them and tell how you would control them.

For the activity:

How much trouble did you have finding fourteen variables that were not controlled? Did you resort to some pretty unlikely ones? You didn't have to. Have you discussed your list of variables with your group? Did all of you have similar lists? You should get some additional ideas from your discussions. The variables listed in the instruction booklet are these:

1. Kind of seed
2. Temperature
3. Number of seeds
4. Depth of vermiculite before adding seed
5. Depth of vermiculite after adding seed
6. Amount of light
7. Air movement
8. Amount of liquid added at start
9. Addition of more liquid
10. Concentration of detergent
11. Kind of detergent
12. Kind of container
13. Number of seeds germinating
14. Type of material in which all seeds are planted

Implications for teaching:

Did this activity give you some insight into the teaching of variables? Could you help your students devise valid experiments? You should think about valid experiments because you will be using experiments in any science program you teach. Does this activity suggest any teaching strategies to you? You might start a lesson with a hypothetical problem or story.

COMMENTS

SUMMARY

An experiment is only as good as its controls. Students learn to identify and control variables that could invalidate an experiment. As an integrated process, this activity builds on the student's ability to use the basic processes.

The student is also preparing for the time when he will have to set up an experiment and defend its validity.

Module 98: Experimenting/c
Eye Power

The last ten modules of SAPA II are experimenting modules and are used to integrate all of the processes. In this module, the students investigate a common phenomenon—the resolving power of the eyes. *Resolving power* is the ability of the eyes to separate an apparent single light source into two or more separate sources. To illustrate this phenomenon, here is an example. At a distance, the headlights of an automobile appear as a single light, but as the automobile gets closer, two distinct lights are seen. As the students investigate this phenomenon, they practice the skills learned in previous SAPA II lessons. Begin this activity to learn about resolving power.[27]

These students are performing Activity B in the Eye Power unit.

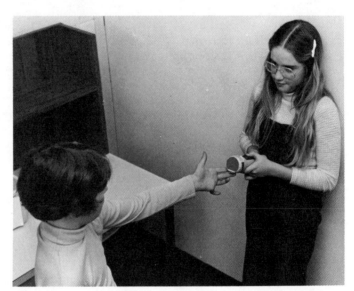

A. How well do you see? Do you always see what you think you see, or do your eyes deceive you? For this exericse, you will need a hand lens, a black and white newspaper picture, and a colored newspaper picture. (Colored pictures in textbooks serve the same purpose.) Try several.

1. Look at a black and white newspaper photograph. Is everything either black or white, or are there shapes of gray in the picture?

How can you print shades of gray using only black ink on white paper?

2. Now, take a hand lens and look at the same picture. What do you see, especially in the gray areas?

   Did your eyes deceive you?

3. Try the same thing with a colored picture. Describe what you see.

4. Why do you see a solid color when you look at the pictures unaided by the hand lens?

5. If possible, look at a television screen with a hand lens (turn it on first). You will find it quite interesting.

By examining a newspaper through a hand lens, this child is learning about the resolving power of his eyes.

B. You will now have the opportunity to see how your eyes resolve two dots of light. You will need a flashlight, a piece of onionskin paper, a construction paper disk, a pin, tape, and a meterstick. You will also need several partners; so do this activity with your group.

Cover the flashlight with the onionskin paper; then tape it in place. Use the pin to punch two holes, about two millimeters apart in the center of the construction paper disk; then tape it to the front of the flashlight, over the onionskin paper. If you have done everything correctly so far, you should be able to see two small light sources when the flashlight is turned on.

Find a test area about six or seven meters long, the darker the better. Have one member of the team stand at a marked spot at one end of the test area and hold the flashlight at eye level. Have another person start at the other end of the area and walk forward until he can see the two light dots. Mark the spot where the dots become distinct; then measure the distance from the light. Repeat the test several times with different people then average the results of each person.

	Team Member	Team Member	Team Member
1. Trial 1	_____	_____	_____
Trial 2	_____	_____	_____
Trial 3	_____	_____	_____
Average	_____	_____	_____

2. How do you account for the discrepancy among different team members?

3. What variables are you manipulating? (You might need to refer to the previous activity for information about variables.)

4. Describe how you manipulate them.

√ SELF-CHECK

For the activity:

Did all of the little dots surprise you? Your eyes cannot resolve anything that small and close together without help. Did you have any trouble with the experiment? If so, were you able to get it straightened out? Did you need to refer to the previous activity *(Module 71: Controlling Variables/c)* for help with variables? If you didn't do that activity, it might be worth your time to look at it. If the instruction booklet is available, you might want to refer to it.

Implications for teaching:

Experimenting is an important part of teaching any science. Did this activity help you understand how to set up an experiment? Have you ever thought of using something simple like a newspaper picture and a hand lens to teach a concept? Sometimes simple equipment is the most effective.

COMMENTS

SUMMARY

All of the basic processes are used when experimenting. The SAPA II modules for experimenting are used at the end of the program to help the child learn to use all of the processes together to really act as a scientist. You can use experiments to help your students understand complex technology as well as everyday phenomena.

Summary: SAPA II

You have reached the end of the SAPA II portion of Part 2-1. The background information provided you with material about the program to help you understand its teaching strategy and how it was developed. Instruction booklets were reviewed so that you could see how each process is developed in small steps and the types of materials and activities that are used in SAPA II. Finally, you were involved in the same types of activities that the students using the program are involved in in order to give you insight into the elementary student's reaction to the program.

Now that you have completed this section of Part 2-1, you should be able to do the following:

1. Identify and describe materials in SAPA II
2. Identify and describe the scope, sequence, and teaching strategy of SAPA II

If you have any questions or doubts about SAPA II, see your instructor. If you feel confident that you can do the tasks, you are ready for the final group seminar and instructor check.

FINAL SEMINAR

Review the activities you have completed for the SAPA II program. Meet with a small group of your classmates to discuss any questions that you have about the program. Be sure to give your reactions to the program. *Be sure to ask your instructor to participate in this final seminar.*

NOTES

BIBLIOGRAPHY

AAAS. *Curriculum Catalog.* (AE9505-10-85) (50MFLPC) Lexington, Mass.: Ginn and Company (Xerox Corporation), 1976.
_____ . *Program Guide* of SCIENCE . . . A PROCESS APPROACH II. Lexington, Mass.: Ginn and Company (Xerox Corporation), 1975.

SUGGESTED READINGS

AAAS. *Program Guide* of SCIENCE . . . A PROCESS APPROACH II. Lexington, Mass.: Ginn and Company (Xerox Corporation), 1975.

Carin, Arthur A., and Sund, Robert B. *Teaching Modern Science.* 2d ed. Columbus, Ohio: Charles E. Merrill Publishing Co., 1975. Pp. 55-59.

Gega, Peter C. *Science in Elementary Education.* 2d ed. New York: John Wiley & Sons, 1970. Pp. 550-73.

Sample Packet, Science . . . A Process Approach II. Lexington, Mass.: Ginn and Company, 1975.

Summary: Laboratory Approach to Elementary Science

In this Part, you were introduced to three of the best-known laboratory approaches to elementary science: *Elementary Science Study* (ESS), *Science Curriculum Improvement Study* (SCIS), and *Science . . . A Process Approach II.* (SAPA II). Background information and hands-on activities were designed to give you a basic knowledge of the program, how it is taught, and what it is trying to teach. You were also asked to see how you could use some of the ideas presented in a classroom of your own. You must be able to draw from these programs (or from any source, for that matter) ideas and strategies that will enhance your own program and improve your teaching ability.

You have worked with small groups and in large groups, sharing ideas. If the only thing that you shared were answers to questions, then you probably have missed the point.

At the beginning of Part 2-1, two objectives were stated. Look at them again and ask yourself if you can do these tasks:

1. Identify and describe the materials in the *Elementary Science Study* (ESS), *Science Curriculum Improvement Study* (SCIS) and its revisions *SCIIS* and SCIS II, and *Science . . . A Process Approach* (SAPA II) programs
2. Identify and describe the scope, sequence, and teaching strategy of ESS, SCIS and its revisions *SCIIS* and SCIS II, and SAPA II

If you think you can achieve these objectives, see your instructor about arranging for a final seminar evaluation. Your instructor will provide you with the necessary instructions.

FINAL SEMINAR

Look at all of the information and notes you have accumulated from your activities, group seminars, and instructor contacts while completing this Part. You are now ready to bring them all together. Your instructor will decide on the appropriate course of action, such as a large group (class) discussion, instructor summary, or small group discussion. Whatever form this synthesis takes, be sure that you participate.

NOTES

COMPETENCY EVALUATION

The background information and activities for Part 2-1 were designed to familiarize you with the identification and utilization of equipment and curriculum materials that can be used to conduct learning experiences for elementary children. You are now ready for the competency evaluation measure for this module if your instructor chooses to administer one. Consult with your instructor for specific directions.

PART 2-2
Textbook Approach
to Elementary Science

FLOWCHART: Textbook Approach to Elementary Science

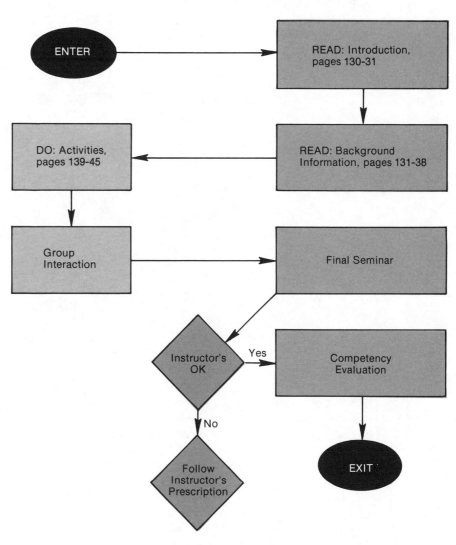

Introduction

The textbook approach is probably the most familiar of the science curricula materials that you will study. Wherever you teach, you probably will have a textbook program to work from. Unfortunately, not all schools with textbook programs have current textbooks in use. Some may be using books that are dated, and others may be using earlier editions of current texts. What will you do with the text, however good or bad, that you have to use?

This Part will cover four areas: (1) science textbooks, (2) reviewing a textbook, (3) activities on reviewing a textbook, and (4) what to do with what you have. Read the "Science Textbooks" section. It will give you information about the textbook approach and how it is used. "Reviewing a Textbook" contains information about review criteria. In the "Activities" section, you will review some textbooks. This will give you the opportunity to examine a book and find out for yourself what actually is in an elementary science textbook. A group discussion of the textbooks reviewed will allow you to share your views with others and to get their opinions of your textbook. You might want to ask your instructor to be a part of this discussion.

Various science textbooks.

You may be asked to make a presentation of some type, based on the textbooks reviewed and your group discussion. Your instructor will provide instruction for this presentation. Do some serious thinking about how you can use what you have learned in this Part and those before it to help you develop and teach a sound science program for your students. Discuss your thoughts and ideas with your group. Meet with your instructor for a competency check or final discussion. This will help you assimilate the information that you have learned.

Once again, the flowchart will be your guide through the Part. Don't forget the group discussions as well as the instructor check. Individual work will provide you with a certain amount of information, but interaction with the group and instructor will broaden your perspective.

GOALS

After completing this Part, you will demonstrate competency in the ability to evaluate various elementary science textbooks.

BEHAVIORAL OBJECTIVES

In completing this Part, you will be able to do the following:
A. Evaluate elementary science textbooks using the criteria presented in this Part,
B. List and explain the strengths and weaknesses of an elementary science textbook program.

Background Information

Textbooks have been around for a long time and are considered by many teachers to be the only way to teach. In this section, you will discover why they are so popular, what their weaknesses are, and how to most effectively use them.

Science Textbooks

In the past, science textbooks were usually science readers. As part of the reading program, children were given a science text and asked to read it. They read about flowers, birds, and a trip to the zoo. Some physical science, such as the planets, magnets, and simple machines, was also included, but generally speaking, nature study was science and went hand in hand with the reading program in the elementary school. Textbooks changed somewhat to include science experiments, or at least they were called "experiments." Too often they were either read about, performed by the

These children will experiment with the science they read about in their textbooks.

teacher as a demonstration, or ignored. Very little actual hands-on experimenting by the students was suggested or carried out. Reading was still the main learning tool. The student was the passive receptor of factual knowledge contained in the textbook.

Current learning theories, however, stress the need for learning experiences to include concrete involvement.[1] This idea has resulted in the formation of new hands-on programs, which are very popular. To compete, textbooks had to revise their approach to include the "concrete involvement" concept. The need for a strong science program, separate from the reading program, was also recognized. Consequently, science textbook series were designed to present a total science curriculum for the school. New philosophies, help for the teacher, provisions for exceptional children (either above or below the norm), and the availability of supplies in the classroom all contributed to the updating of modern elementary science textbooks. With all the new programs available, why do so many schools still use traditional textbooks? A good question, with several good answers:

1. Good textbooks are excellent teaching aids. They are a resource for the teacher, providing factual material, experiments, demonstrations, illustrations, and references as well as suggestions for teaching. For the students, they provide information about specific subjects or topics, suggest activities or experiments, and provide illustrations that might not be available otherwise. Textbooks are primarily designed to be sourcebooks for teachers and students, and to provide everything needed to successfully complete a specific amount of work.

2. A textbook series provides the school system with an organized science program. Most schools do not have the expertise, personnel, or financial resources to develop a K-6 science curriculum. Commercial publishers, however, can develop a sound program with professionals who have experience in curricula, elementary science, and child development. The school system knows that by adopting a particular science series, it will have a balanced, sequential science curriculum that will provide a basic science education for children.

3. Textbooks provide teachers with organized units of work. Most teachers like to teach by the unit method in which one topic is explored thoroughly over a period of time, using a variety of activities. The scope and sequence of each topic are clearly defined for the teacher. He knows exactly what to cover without infringing on material covered in other grade levels.

4. Textbooks are especially helpful for beginning teachers. The teacher's manual provides general suggestions for teaching the unit as well as specific suggestions, page by page, for teaching the material to be covered. The beginning teacher can find security in knowing exactly what is to be taught and having directions for teaching it.

5. Textbooks may be less expensive to purchase and use than some of the kit or text-kit materials. It depends upon the school system and the way that it purchases, uses, and supports the program that is adopted. If cost is really going to be a significant factor in the type of program that will be selected and used, a detailed cost analysis of all programs and textbooks under consideration should be made.

6. Textbooks are popular in many schools because the teachers and administration do not know that any other programs exist. In this situation, you will probably

[1]For more information, see works by Piaget, Bruner, and Gagné in the Suggested Reading for this Part.

find that textbooks are selected by the administration from a small list of publishing companies who have been supplying similar books in the past. Wherever you teach, you should try to provide information about new materials to your colleagues.

7. One last reason for the popularity of textbooks with some teachers is that they feel insecure with any other program. This is not a good reason but, unfortunately, is a very valid one for selecting, not just the textbook approach, but a specific textbook series. Many teachers are unsure of themselves, especially in science, and need the security that the textbook can provide. It tells the teacher exactly what to say and do, what to expect the students to do, what questions to ask, and specific answers to questions. Teachers are never put in the position of being asked questions they cannot answer or of having to decide what should be taught next. Everything is in the book.

In choosing a science program, teachers should realize that the *best program* is a relative term. What might be the best choice for one school might be the poorest choice for another. Each school must consider all possible programs and select the one that suits its particular needs. Only then can the school claim to have selected the best program available.

There are some inherent weaknesses in textbook programs which should be considered. Think about how you might be able to overcome these weaknesses if you use a textbook:

1. Too often the readability of the textbook is below or beyond the level of the children using it. If children cannot, or will not, read the textbook, obviously they cannot get very much out of it. Most publishers are trying to control the reading level of their textbooks to make them more attractive to students.

2. Many teachers use the textbook as the only source of information. However, the textbook cannot include everything that any student needs or wants to know about a subject. Teachers need to supplement textbook activities just as they supplement nontextbook activities. Quite often, the teachers who limit themselves to the textbook are either a little bit lazy or they are very insecure and are afraid to do anything not found in the textbook. Publishers can, and do, include suggestions for supplementary work, but they cannot force the teacher to use the suggestions.

3. Textbooks are generally designed to be read. They try to substitute words for hands-on experiences. This is a serious weakness if you consider the work of Piaget, Gagné, Bruner, and others who have been mentioned throughout this text. Teachers should try to involve the students as much as possible in the activities to help provide hands-on opportunities that go along with the reading activities. Furthermore, many children cannot read or are severely limited in their reading ability. They are heavily penalized if their work is based totally on textbook material.

4. Textbooks tell all the answers. Experiments and illustrations are shown step by step, from beginning to end. The children follow along to the end where they are told the answer. There is no opportunity for students to question, make suggestions, or find out answers. They are passive receptors, not actively involved learners.

5. Each textbook series has developed its own scope and sequence, which creates a curriculum problem. There is always a question as to the proper grade placement of the various science concepts and principles, as well as which ones should be taught or omitted. Since textbook series do not agree on this subject, the schools have to find the textbook that best agrees with their scope and sequence philosophies. If no textbook meets the requirements of a school, then the school must modify its program to fit the text, rather than modifying the textbook to fit the school.

6. One last weakness of the textbook approach is not really the fault of the textbook but of the users. Textbooks become outdated. Publishers revise them periodically and bring out new editions for the market; however, sometimes publishers will only face-lift the programs and not update the content. This doesn't help teachers stay on top of what's new in elementary science. Probably more serious though is that too many schools do not update their textbook series. A series is adopted, and it then remains until the covers fall off. It is hard to imagine that a teacher must teach from a textbook that is over twenty years old, but it is done. Although it is hard to get rid of a book that is still in good shape, it must be done periodically to keep from using outdated programs.

As you can see, the textbook approach has strengths and weaknesses. Many teachers think that the strengths far outweigh the weaknesses, especially if you know the weaknesses and can compensate for them. Is the textbook approach for you? This question can only be answered by you when you start teaching. But for now, think about what the textbook has to offer you as a teaching aid and how you can use it in your classroom.

There are several ways to use the textbook effectively. Remember that it is a teaching tool, not the entire science program. There are many ways of using a textbook for effective teaching. Your first task is to decide (a) whether or not you like your textbook, and (b) how much you intend to use it. These two decisions go hand in hand. If you like your text and consider it suitable, you will use it much more than if you consider it to be a poor text. There are four ways for a teacher to use a textbook in her classroom.

1. It is not used at all, except to occupy space on the bookshelf. Obviously, this is a poor use of the textbook. This situation might occur for two reasons. First, the teacher doesn't like science or is afraid to try it; so she doesn't teach it. There is no excuse for this. The science program is as important as any of the other programs. Second, the text is either so obsolete or so poorly structured (in the teacher's eyes) that it offers no real help to the teacher. If this is the case, then the teacher probably has developed her own science program and is teaching it with supplementary materials and trade books.

2. The textbook is used occasionally as a reference. Usually this situation arises when the teacher considers the text inadequate for some reason and develops his own program. This is a very inefficient use of the textbook, and the teacher might be just as well off without it.

3. The textbook is used as a resource, supplying ideas for discussion, demonstration, and experiments. It also furnishes background information about the topic being studied and verification of experimental results. This is probably the best way to use a textbook. A partnership exists between the teacher and the textbook—the textbook helps the teacher to teach.

4. The teacher relies very heavily on the textbook, following it very closely. In extreme cases, the textbook may literally be used as a science reader. Usually though, the teacher follows the teacher's guide step by step, suggestion by suggestion, page by page from cover to cover. This overdependence severely limits the teacher and students, whether they know it or not. A majority of teachers are probably teaching this way.

How should you use the textbook? As a beginning teacher, you will probably rely on the textbook more than you should, and there are good reasons for doing so. The textbook will provide you with a course outline, teaching suggestions, content information, appropriate grade level material, and suggestions for evaluating the students. All of this you need to know but probably do not because of your lack of experience. The textbook can give you the security you need to do a good job, because you know that if you follow the textbook, you will teach what you are expected to teach and will achieve reasonably acceptable results.

After your first few lessons, you should be confident enough to begin doing more than merely following the book. Continue to use the textbook as a guide, but also begin to use some of the other strategies you have learned as a supplement to your basic approach. Try to move to step 3 above. Use the textbook as an aid to help you teach but don't be limited by it.

Try using the textbook as a source for experiments, demonstrations, or discussion topics. Read the material, then involve the students in discussions or experiments before they read the material. Then they can go to the textbook and see how their experiments compare with those given. This strategy keeps the textbook from giving the answers before the children ever have a chance to try to find them out. You can also use the textbook to provide data or background information for your topic. The material is in the book, so have the children read it. Don't be afraid to bring in supplemental material or to modify the textbook materials to fit what you are trying to do. Remember that you are the teacher and the textbook is an aid. The text cannot teach for you, although many teachers seem to think it can.

Reviewing A Textbook

Suppose that you have just been given a copy of the textbook you will be using next year or that you are at a book display and see a new science textbook. What are you going to do? You will probably look through the book, but do you really know what you are looking for? If you do, you are indeed a rare individual, because most teachers don't. In this section, you will be given some guidelines to help you evaluate or review a textbook, then you will have the opportunity to actually review one of the science textbooks. The following material will discuss eight items to consider when reviewing or evaluating a textbook.

AUTHOR(S) AND PUBLISHER

Who wrote and published the textbook? This might seem minor, but it is not. Does the publisher have a good reputation in the textbook field? There are some good materials available from minor publishers, but most of the established programs come from the major publishers. Who are the authors and what are their credentials?

These preservice teachers are trying to develop material to supplement their textbook.

Is this their first series, or have they written others? The answers to these questions can give you a clue as to what to expect when you look through the text. You may not know or recognize the names of the author, but you should recognize names of the major publishers and be familiar with the types of material associated with each of them.

APPEARANCE

This is probably the first thing you will notice, and it is important. It is very difficult to get a child interested in a dull-looking book, no matter how well written it may be. Eye appeal does play a major role in the acceptability of a textbook to the teacher and students who have to use it. Look for such things as the use of color, illustrations (we will come back to this point later), format, and the physical size of the textbook. A book that is too large or too small is awkward to handle and detracts from the usability of the book. Look at the cover and paper. How durable will the book be? Does it look like a book that you might like to use?

ILLUSTRATIONS

Illustrations serve two purposes in a textbook: (1) to add to the overall appearance and (2) to illustrate the material. Good illustrations, either photographs or drawings, brighten up the book by adding to the black and white of the print and paper. Are the photos current or dated? Posed or realistic? Proper placement can also make the reading material seem a little less formidable to the slower readers. The primary purpose of an illustration is, of course, to illustrate some specific material. Students can read the materials and look at the picture. They might also "read" the picture to help them understand the material they have read. This, again, is a great help for the slow reader. As you look at the illustrations and/or photos in a textbook, think about what they do for the book. Do they merely decorate, or do they also teach? Do they make you want to look further, or do they turn you away from the book? Would your students like the illustrations?

READABILITY

There is more to readability than proper vocabulary. First of all, look at the size of the type. Is it too small, too large, or just right to be easily read? Does the layout of the reading material allow you to easily find the beginning and then to follow through without losing your place as you read? Is the reading material appropriate to the children using it? Too often, authors seem to use vocabulary and writing styles that are more appropriate for upper-level students than for elementary students. The textbook is not really of much value if your students cannot read it.

CONTENT

This is the material that you will be teaching; so you should look at it closely. Is it scientifically accurate? You would think that it would be but sometimes errors do appear either through oversight or because the information is obsolete. A good textbook should contain current, accurate information. Who wants to read that people may someday go to the moon when you know that they have already been there? If the textbook is obsolete or inaccurate, you might want to try to find a different text. Does the content fit the students using it? If it is too simple or too advanced, you will have trouble getting your students to relate to it. Are the analogies appropriate? Analogies and examples should relate to the child's world, not the teacher's, since the student is the one who is trying to understand the concepts being illustrated. How is the content organized? It should be in a logical sequence that children can follow. Look to see if you can determine the pattern, or teaching strategy, of each lesson or unit. Can you find the central theme? This is the foundation of the textbook, and you usually find it explained in the teacher's manual, but you may have to look for it. If no central theme exists, the book may not have very much continuity. Does the textbook include any materials about, or related to, any of the various cultures of the world? Science is not a cold, impersonal field. The effect of science on culture should be included to help students understand how science can benefit, or harm, humanity. How does the textbook treat controversial materials? There always have been and always will be differing opinions in science. Does the textbook ignore controversy, or present it for discussion? See if you can find suggested ways of teaching controversial material.

The content of the textbook is important to you and to your students because it is the reason for the textbook's existence. Everything else (teaching strategies, experiments, illustrations, etc.) is directly related to the effective teaching of the content found in the textbook in this approach to elementary science education.

PRESENTATION OF MATERIALS

There must be a method for presenting the content material to the student. Each textbook will have its own method or teaching strategy. Can you describe the teaching strategy in your textbook? What kinds of questions are asked of the student? Look for memory-based as well as thought-provoking questions. How are the students involved in the learning process? Are experiments conducted by the students, or are they demonstrated by the teacher? How much reading is required of the student? Find out if your textbook is a *reader* or a *guide* for the students. What does the teacher do in your program? Examine the teacher's role and try to determine how you would teach the material. Would you have much freedom, or is the teaching procedure highly structured? As you look at a textbook, you should remember that not only do the students have to learn from it, you have to teach from it. If the material is not presented in an interesting manner or does not require much thought or effort from the students, does it really matter what content is being taught or what the pictures look like?

MATHEMATICS

You are not going to teach mathematics using your science textbook, but some science does involve the use of mathematics. Mathematics can be called the language of science, and children need to know the basic mathematics that will be needed to carry out the activities in the textbook. Does your textbook discuss or use mathematics? What mathematics skills are necessary to do the science activities? Do the children have these skills, or will you have to teach them? You may have to correlate your science and mathematics lessons to be successful.

TEACHING AIDS

How much help does the textbook give teachers? Look for a teacher's manual or a teacher's edition of the text. How is it organized? It should give you some background information about the philosophy, goals, objectives, and teaching strategy followed by a reproduction of the actual text with suggestions for teaching the material. Is the teacher's manual usable? If it is too large and clumsy or the print is too small or you have trouble finding the teaching suggestions, then you may need help that this teacher's edition cannot give you.

How complete are the teaching suggestions given? Some manuals offer only general ideas, but others provide specific directions. Would you feel confident using the teaching suggestions provided, or would you want more? Are references provided? A listing of supplementary reading material for both teacher and students can be very helpful, especially for beginning teachers who have not had the opportunity to develop their own lists. Does the manual suggest any films, filmstrips, recordings, or sets of pictures that might be used with the textbook? You should supplement textbooks with audiovisual material. Remember, one weakness of textbooks is that they tend to be used as the only source of information. A good list of supplementary materials, with suggestions for using them, can help the teacher overcome this weakness.

Now you should have some information to help you judge a textbook. There are other considerations you will probably include in your review that will reflect your own concerns and prejudices. This is why people, using the same list of criteria, will rate textbooks differently. The final judgment on the appropriateness of any textbook is based on personal opinion.

Reviewing a Textbook

The purpose of this activity is twofold. First, you will apply what you have learned by actually reviewing a textbook, and second, you will become familiar with one or more of the textbooks available to you. Remember to think about how your knowledge of textbooks can help you become a better teacher.

Group discussions and instructor checks will be an integral part of this activity, and you may be asked to make a presentation to the class. Your instructor will help you decide on the appropriate type of presentation. You will share information about several textbooks or textbook series as you interact with your group, class, and instructor.

Access to several science textbook series will be necessary. Your instructor will direct you to the textbooks available for your use. Here is a list of textbook series and the publishers to get you started. This is not a complete list, as it includes only current, major publications. There are other series that might be locally available. Two programs are separated from the others. Technically, they are textbook programs but they have special characteristics which will be explored in Part 2-3. Do *not* use them in your review unless directed to do so by your instructor.

Publisher	Title	Copyright
Cambridge Book Co.	Cambridge Work-a-Texts in Science	1973
Charles E. Merrill Publishing Co.	Discovering Science Series	1970, 1973
D. C. Heath & Co.	Heath Elementary Science	1973
Ginn and Co.	Ginn Science Program	1975, 1977
Harcourt Brace Jovanovich	Concepts in Science (3rd ed.)	1975
J. B. Lippincott Co.	The Elementary School Science Program	1977
Laidlaw Brothers	The Laidlaw Exploring Science Program	1976
The Macmillan Co.	Learning Science	1977
Rand McNally & Co.	Elementary Science: Learning by Investigating (ESLI)	1973
Silver Burdett Co.	Science: Understanding Your Environment (S:UYE)	1975
McGraw-Hill Book Co. (Webster Division)	Gateways to Science	1979

These two programs will be discussed in Part 2-3.

Publisher	Title	Copyright
Addison-Wesley Publishing Co.	Space, Time, Energy, Matter (STEM) Elementary School Science	1975
Houghton Mifflin Co.	Modular Activities Program in Science (M.A.P.S.)	1974, 1975, 1977 (Revised 1979)

A. Select a textbook or textbook series to evaluate. Use the teacher's edition if possible. Your instructor will work with you to help you decide which of the following suggestions to use:

1. Select a textbook or textbook series of your own choosing and proceed with the textbook review in this section.
2. Meet with your group and decide how to proceed. A particular series might be selected so each member of your group could review a different grade level. Or, one grade level might be chosen so each group member could review a textbook for that grade from different series.
3. Follow your instructor's directions.

B. Start your review. Think about the suggestions that were presented earlier (pages 134-38). You might want to refer to them as you go through the textbook review activities.

Textbook Review

Overall Textbook

Publisher:

Title and Grade Level:

1. Examine the front matter (preface, introduction, contents, etc.) of the textbook.
   a. What is the copyright date?

   b. What are (or seem to be) the overall goals of the textbook?

   c. Describe and give examples of the organization of the textbook. (Is it organized in single lessons? Topic form? Units? etc.)

2. Examine several specific lessons.
   a. Describe the typical lesson format.

b. Describe at least two specific teaching techniques suggested.

c. Identify the main kinds of materials suggested for use with the lessons.

d. What kinds of objectives are indicated for the lessons? Give examples.

e. Describe evaluation techniques suggested.

f. Your comments. (Any unique features? etc.)

Specific Criteria

Rate your textbook using the following rating scale and criteria. Space is provided for the other members of your group to record the ratings of their textbooks. This should allow you to compare several textbooks. Think about your ratings; not every item will be a 5, even if you really like the text.

Rating Scale

0—Book totally lacking in the characteristic
1—Occasional evidence of the characteristic
2—Evidence of the characteristic but below average
3—Frequent evidence of the characteristic
4—Excellent evidence of the characteristic
5—Superior in all aspects of the characteristic[2]

---

[2]Robert B. Sund and Leslie W. Trowbridge, *Teaching Science by Inquiry in the Secondary School* (Columbus, Ohio: Charles E. Merrill Publishing Co., 1973), p. 451.

Names of Series Reviewed

CRITERIA FOR RATING TEXTBOOK

Appearance					
1. Attractive and appealing to children					
2. Margins and page arrangements contribute to readability and attractiveness					
3. Adequate spacing and appropriate type size					
4. Suitable size for easy handling					
5. Durable backings					
6. Good quality paper					
Illustrations					
1. Contribute to meaningfulness of the content					
2. Interesting and scientifically accurate					
3. Clearly produced and well placed on the page					
4. Placed near the text they illustrate					
5. Appropriate to the grade level					
6. Clear in meaning					
7. Current and attractive					
Readability					
1. Reading level appropriate for the children using the book					
2. Appropriate page layout—students can easily follow the sentences					
3. Type size and style easy to read					
Content					
1. Develops problem-solving skills					
2. Scientifically accurate and up-to-date					
3. Appropriate for the developmental level of student					
4. Analogies and activities appropriate					
5. Follows a logical sequence					
6. Develops positive attitudes toward science					
7. Suggested lessons stimulate interest that will lead to further study					

	Names of Series Reviewed					

CRITERIA FOR RATING TEXTBOOK (continued)

8. Provides for various ability levels (the nonverbal child, the child with reading problems, the high achiever, etc.)						
9. Well balanced in terms of scientific content						
10. Central theme clearly defined						
11. Free from anthropomorphism, teleology, and personification (attributing human form, qualities, purpose, or will to nonhuman things)						
12. Usable index and table of contents						
13. Glossary of science terms with clearly stated explanations of meanings						
14. Free from sex bias						
15. Suggestions for treatment of controversial materials						
Presentation of Material 1. Has a clearly defined teaching strategy						
2. Uses class discussion						
3. Uses divergent as well as convergent questioning						
4. Encourages children to do experiments						
5. Material presented in an interesting manner						
Mathematics 1. Includes mathematics in the experiments and activities						
2. Offers suggestions for using mathematics						
Teacher's Guide 1. Offers alternate activities						
2. Gives necessary background information for effective use of textbook material						
3. Suggests teaching aids, games, etc.						
4. Lists resources (printed, visual, audiovisual, etc.)						
5. Suggests remedial and/or enrichment activities						
6. Suggests a variety of evaluation techniques						
7. Offers help in planning and implementing text material						
Total points for textbook (225 possible)						

C. By now, you should have some definite opinions about the textbook you have been reviewing.

1. What do you consider to be the *strengths* of the textbook?

2. What do you consider to be the *weaknesses* of the textbook?

3. Would you like to use this textbook in your classroom? Why or why not?

D. Meet with your group and discuss the textbooks you have reviewed. Record some of your comments and discussions.

E. Arrange for a group discussion with your instructor. Together you will make plans for an appropriate competency evaluation for this Part. Some suggestions are offered below:

1. One member of the group will act as a spokesperson to present a series to the class, using the criteria on the rating sheet to prepare the report but condensing it to a usable format.
2. Each member of the class will make a presentation to his group and instructor.
3. The instructor will make suggestions or requirements for the evaluation.
4. Design your own evaluation and submit it to your instructor.

√ SELF-CHECK

For the activity:

Did you have any trouble finding the information you were looking for? If so, that might indicate a weak area in the textbook. Did you rate everything high because you liked the textbook or low because you didn't like it, or did you try to overcome your biases and be honest? Sometimes this is hard to do, but you have to try.

Implications for teaching:

Do you have a clear concept of the textbook approach to teaching? You will probably use textbooks when you start teaching; so you should be especially concerned about using them. Did you find out what kinds of activities are considered appropriate for particular grade levels? Think about this, because it can be important. Textbooks can give you, the beginning teacher, information about the capabilities of students at different age levels and about the topics generally covered in specific grade levels. This might be a source of help when you plan your own science program.

COMMENTS

BIBLIOGRAPHY

Sund, Robert, and Trowbridge, Leslie, *Teaching Science by Inquiry in the Secondary School.* 2d ed. Columbus, Ohio: Charles E. Merrill Publishing Co., 1973.

SUGGESTED READING

Bruner, Jerome S. *Toward a Theory of Instruction.* Cambridge: Harvard University Press, 1967.

Carin, Arthur, and Sund, Robert. *Teaching Modern Science.* 1st ed. Columbus, Ohio: Charles E. Merrill Publishing Co., 1964.

Collette, Alfred. *Science Teaching in the Secondary School.* Boston: Allyn and Bacon, 1973.

Gagne, Robert M. *The Conditions of Learning.* New York: Holt, Rinehart and Winston, 1965.

Piaget, Jean. *Six Psychological Studies.* New York: Random House, Vintage Books, 1968.

Piltz, Alfred, and Sund, Robert. *Creative Teaching of Science in the Elementary School.* 2d ed. Boston: Allyn and Bacon, 1974.

Summary: Textbook Approach to Elementary Science

In this Part, you have been introduced to the elementary science textbook. The background information provided you with some of the strengths and weaknesses of textbooks as well as suggested ways of teaching from a textbook. You were given some suggestions as to what to look for when reviewing a textbook, and a list of criteria was also presented for your consideration. These suggestions and criteria were used when you actually reviewed a textbook in the activities section.

The purpose of this Part was to make you familiar with the most widely used of all science programs, the textbook. The objectives were that in completing this Part you will do the following:

1. Evaluate elementary science textbooks using the criteria presented in this Part
2. List and explain the strengths and weaknesses of an elementary science textbook program

Do you think you have met the objectives? If so, see your instructor for the final seminar. If not, go back over the material, then see your instructor for help.

FINAL SEMINAR

Review your textbook evaluation and the background material you have read. Meet with your small group and be prepared to bring up any unanswered questions that you have and to help answer those that your classmates might have. *Your instructor might want to participate in this seminar so be sure that he is invited.*

NOTES

COMPETENCY EVALUATION

You should now have some knowledge of textbooks and their use in the classroom. Your instructor may choose to involve you in some type of competency evaluation over the material you have studied. Check with your instructor.

PART 2-3

Text-kit Approach
to Elementary Science

FLOWCHART: Text-Kit Approach to Elementary Science

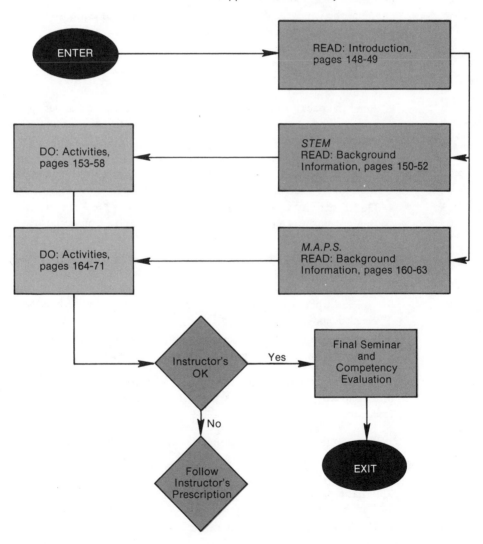

Introduction

In this Part, you will be introduced to two science programs that attempt to combine the best features of the science textbook and the laboratory kit. The programs are these:

1. *Elementary School Science: Space, Time, Energy, and Matter* (STEM)
2. *Modular Activities Program in Science* (M.A.P.S.)

The pendulum continues to swing in education. It is presently in a neutral position with respect to science education. We know that textbooks are not all bad and laboratory kits are not all good. Each has strengths and weaknesses. This knowledge has allowed the development of a new type of science program: the text-kit approach to elementary science. The two programs examined here are basically textbook programs, based on the laboratory approach. They try to combine the strengths of both approaches, thus eliminating some of the weaknesses. You will have to be the judge of how well they succeed.

This is the sequence to follow in this Part:

1. Do each section separately. Complete one before going on to the next so that you do not confuse them.
2. Read the background material first. This will help you understand the philosophy and methodology of the program before you start reviewing the teacher's manual or the textbook.
3. Examine the teacher's manual and textbooks for the program. You might want to examine manuals from several grade levels, but first examine one thoroughly, then look at some of the others.
4. Do the activity or activities. This will give you a "feel" for the materials used and the teaching strategy involved. Think about how you could use the activity, or a similar one, in your classroom.
5. As you complete each activity, use the self-check to help you put everything into its proper perspective. After you have reviewed both programs, meet with your instructor for an instructor check. You should talk with your instructor any time you need help, not just at the end of the Part or an activity when an instructor check is indicated.
6. Participate in a competency appraisal at the end of the Part. This will help you evaluate your knowledge of the programs presented.

If you follow the sequence outlined, you will not have any trouble achieving the goals and objectives of this Part. Here is some specific information that you should look for:

1. How do these textbooks compare with the traditional textbooks reviewed in Part 2-2?
2. Can you find examples of process science included in the programs?
3. What types of laboratory materials are used? Could you obtain similar equipment for your classroom?
4. Could you adapt some of these activities for your classroom if you had a different program?

5. Would you like to teach any of these programs? What do you consider the strengths and weaknesses of each program?

This list is by no means complete, and you should add your concerns to it. By now, you should be familiar enough with science programs to begin developing your own personal criteria. Don't close your mind to new, old, or different materials, but do evaluate them on the basis of appropriateness for you and your personal philosophy of teaching.

The flowchart will be your guide to successful completion of this Part. Follow it, and if you have any questions, see your instructor.

GOALS

After completing this Part, you will demonstrate competency in the identification and utilization of science text kit materials that can be used to conduct learning experiences with elementary school students.

BEHAVIORAL OBJECTIVES

In completing this Part, you will do the following:

A. Identify and describe the materials used by these programs:

1. *Elementary School Science: Space, Time, Energy and Matter* (STEM)
2. *Modular Activities Program in Science* (M.A.P.S.)

B. Identify and describe the scope, sequence, and teaching strategy of these programs:

1. *Elementary School Science: Space, Time, Energy and Matter* (STEM)
2. *Modular Activities Program in Science* (M.A.P.S.)

Elementary School Science
Space, Time, Energy, & Matter (STEM)

Background Information

Authors:	Verne N. Rockcastle
	Frank R. Salamon
	Victor E. Schmidt
	Betty J. McKnight
Publisher:	Addison-Wesley Publishing Company
	Menlo Park, California

The STEM program is basically a textbook series for K-6, augmented by the addition of hands-on activities and laboratory experiences. Four major content themes are developed in the series: space, time, energy, and matter. The use of the science processes is also taught throughout the program. Students work alone or in groups, using their natural curiosity. The teacher and the textbook are used to provide support and direction for the student.

Various items in the STEM program.

There are two basic beliefs underlying the STEM program. First is the belief that the intellectual development of the learner and the nature of science are closely related. One cannot be emphasized at the expense of the other. For this reason, equal emphasis is given to developing process skills and to teaching conceptual schemes. The second belief is that the best way for children to learn is through frequent social interaction and direct personal experiences. The sharing of ideas and cooperation in doing the activities are encouraged.

Features of the Program

There are several features of the STEM program that combine to make the program unique. The most important feature is that the program is activity oriented. Students are involved in hands-on activities using common, everyday materials that can usually be found around the home. The use of elaborate materials is discouraged because they are not readily available to the students except in the classroom, and students are encouraged to repeat the activities outside the classroom to reinforce their learning.

Another feature of the program is the behavioral objectives provided for the teacher at the beginning of each lesson. Students should change as a result of their learning experiences, and the teacher should know what kind of changes to expect. Objectives guide the teacher in doing this.

STEM provides for the individual differences of the students. Most students can do the work suggested in the textbook, but exceptional (above average) children can be given additional work tailored to their particular needs. Suggestions and activities are provided in the teacher's editions.

Furthermore, there is controlled readability at each grade level. Available research was used to keep the reading level consistent with the ability of the students using the material. Illustrations are used throughout the program to help the slower reader. Some activities are so fully illustrated that little or no reading is required, minimizing the students' dependence on the textbook and their reading skills.

At the beginning of each teacher's edition, a science vocabulary of new words is given. The number of the page where each word is introduced is also given. Most of the vocabulary used throughout the program is the same as that used by the students in their everyday speech and correlates with the vocabulary found in most of the basal readers.

A scope and sequence chart in each teacher's manual provides the teacher with an overview of the entire K-6 program.

Organization for Teaching

The teacher's manual provides suggestions for teaching daily lessons using the following format:

1. *Concepts:* The concepts to be developed in the lesson are explained to the teacher.
2. *Behavioral objectives:* Specific behavioral objectives for lessons are given.
3. *New words:* A list of new words that will be introduced is provided.
4. *Presenting the lesson:* A list of the materials needed for the lesson is given, followed by specific suggestions for reviewing the written material on each page. Activities are suggested to supplement the teacher's discussion. As an integral part of each lesson, activities are presented to the student as "Something to Try." Suggestions for using these activities are also given.
5. *Evaluating the learning:* At the end of each lesson the teacher is told how to evaluate the lesson and what type of performance to look for. He is expected to adjust the program to his teaching style as needed. New teachers might want to follow the suggestions more closely until they have more confidence in their teaching ability and are used to the program.

Materials

Probably the one feature that separates STEM from a regular textbook program is the materials used. Common everyday materials are the backbone of the program. Most, if not all, of the materials can be brought from home by the students or should already be available at school. Addison-Wesley does provide STEM kits, but they contain only materials that are harder to get, such as hand lenses, balances, and magnets. Materials for activities are necessary for the success of the STEM program, that is, if the students are to learn by doing.

If you have read the background information, you are ready to do the activities from the STEM program. There will be two activities for you to complete. The first activity will be to review a teacher's edition for the program. The second activity will involve you in a sample lesson. From this, you should develop an understanding of the program, as well as be able to add new ideas for teaching to your own repertoire.

Activity 1: Review of STEM Teacher's Edition

A. As you review one of the STEM manuals, look for the following:

1. Suggested instructional techniques and methods
2. The type of equipment used and the way in which hands-on activities can be incorporated into a textbook program

Obtain one of the STEM teacher's editions for review. Check with your group to see what grade levels the others are using. Try to have a representative sample from several levels.

B. Start at the front of the teacher's edition, making a note of the grade level. Look especially for the following information:

1. The foreword and the section entitled "The Addison-Wesley Elementary School Science Program." Read this material carefully.
2. The "Scope and Sequence" chart. Can you see how these four fundamental concepts are developed?
3. The section entitled "About This Teacher's Edition." Read this carefully, especially the part about the teaching strategy.
4. Turn through the rest of the manual, and look at the way the material is presented. Read one lesson.
5. Look at the student page as well as the teacher's instruction.
6. Think about how you could use some of these ideas in your teaching. Does this look like a program you would like to teach?

Components of STEM kits.

C. Make some comments about your review. These comments should reflect your likes and dislikes as well as general statements about the program. These notes are to help you remember what was in the manual.

D. Have a group discussion about the STEM program. Discuss strengths, weaknesses, and personal opinions. Also, talk about how you could use some of the ideas presented. Make notes about your discussion.

COMMENTS

√ SELF-CHECK

Did you really read the introductory material for the teacher, or did you skim over it? This is a good place to find out what the authors are trying to do with the program. Did you go through one of the lessons? If so, you can see how the experiments, activities, reading, and thinking all come together for learning. Did you invite your instructor to participate in your group discussion? Don't forget, she might enjoy participating in your discussions and might add something to your observations.

COMMENTS

SUMMARY

The teacher's manual tells you about the design and philosophy of the program as well as how to teach it. You can also see a copy of the pages from the student's textbook. This activity should give you insight into the STEM program.

Activity 2: Activity for STEM

This activity is adapted from a fourth-level lesson entitled *Heat From Mechanical Energy.*[1] It was selected because simple equipment and familiar occurrences are used to teach a more difficult concept. Do the activity with the other members of your group.

Think about the activity as you go through it. It is quite simple, and you may have had some or all of the experiences already. That is all right because you should be thinking about familiar experiences to incorporate into your

[1]Adapted from *STEM Elementary School Science* by Verne N. Rockcastle, Frank R. Salamon, Victor E. Schmidt, and Betty J. McKnight. Copyright © 1977, 1975, 1972 by Addison-Wesley Publishing Company, Inc. Reprinted by permission.

own teaching program. Simple, easy-to-understand activities can teach much more than can elaborate, difficult ones. Can you think of similar, simple tasks to help explain the concept of heat from mechanical energy?

Heat from Mechanical Energy

A. *Heat from friction.* What happens to the temperature of two objects when they are rubbed together? You will need a partner, a measuring tape and an object about 1 or 2 cm in diameter. A spool or dowel rod would serve. STEM suggests using an empty adhesive tape spool. Hold the round object as shown, with the tape around it.

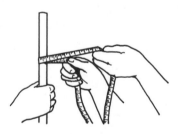

1. Have your partner pull the measuring tape back and forth around the round object several times. What happens to the object as the tape rubs against it?

What would happen if you used your finger for the object?

2. Let the object cool, then change places with your partner and try again. Can you make the object too hot to hold?

Do you think that you could make it hot enough to start a fire? Why or why not?

3. While the object is still hot, compare its temperature with that of the tape. Which feels warmer? Why? Can you explain it?

4. You have obtained heat from friction. Can you think of other examples that you might use in your classroom?

B. *Heat from collision.* What happens to the temperature of objects when they collide? To find out, you will need some colliding objects. Obtain a hammer and a block of wood.

1. First, check the temperature of the block of wood by holding it to your upper lip or cheek. Your fingers are not as sensitive to temperature changes as these areas are. You might try other areas of your body to see where you can best detect temperature change.

2. Put the block of wood on something solid, like a concrete floor. *Do not use desks for this activity.* Now, pound it ten times with the hammer (hit it in the same place each time). Immediately after the pounding is completed, check the temperature of the wood at the place it was pounded. What did you observe?

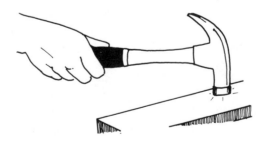

3. Do it again. What happens as the hammer and wood collide more and more often?

4. Suppose you observed no temperature change until you had pounded the wood twenty times. Does this mean that there was no change after you hit the block only once? Why?

C. *Heat from bending.* What happens to the temperature of an object when it is bent? You will need a paper clip to find out.

1. Open up your paper clip. Then bend it back and forth a few times. Touch it to your lip. What do you observe?

2. Bend it a few more times and check the temperature again. Does the number of times the paper clip is bent affect the temperature?

   Does the speed of the bending affect the temperature?

3. Suppose you have a piece of plastic, such as a straw, and you bend it. Will there be a temperature change like the one found in the paper clip? Explain your answer.

SELF-CHECK √

For the activity:

Did you observe temperature changes in all of the activities? You should have. If not, try it again. Where is the best place on your body for you to observe a temperature change? How accurate is this means of detecting temperature change? Can you think of a better way?

Implications for teaching:

Can you think of similar examples to use in your classroom? How could you enrich this activity for a faster student? Think about your faster students as well as your slower ones and have something for everyone.

COMMENTS

SUMMARY

Simple activities are used to teach the child about heat—what it is and what causes it. STEM activities provide the student with factual materials and "Something to Try" activities on almost every page.

Summary: STEM

The background information and the activities provided you with information about the rationale of the STEM program, teaching strategies, unique features, and evaluation procedures. The activities showed you the types of exercises used in the program, as well as a sampling of the type of materials used.

The behavioral objectives for this portion were that in completing it you will do the following:

1. Identify and describe the materials used by the *Elementary School Science: Space, Time, Energy, and Matter* (STEM)
2. Identify and describe the scope, sequence, and teaching strategy of STEM

You should be able to do this now. If not, see your instructor for additional help. If you think you are ready, arrange for the final seminar with your instructor.

FINAL SEMINAR

Look back over the material before you meet with a small group of classmates to discuss the STEM program. Make a note of any questions or comments you want to bring up in the meeting. *Ask your instructor to participate in the seminar.*

NOTES

BIBLIOGRAPHY

Rockcastle, Verne N. et al. *STEM Elementary School Science.* Teacher's Editions, all levels. Menlo Park, Calif.: Addison-Wesley Publishing Co., 1975.

SUGGESTED READING

Rockcastle, Verne N. et al. *STEM Elementary School Science.* Teacher's Editions, all levels. Menlo Park, Calif.: Addison-Wesley Publishing Co., 1975.

Modular Activities Program in Science (M.A.P.S.)

Background Information

Authors: Carl F. Berger
               Glen D. Berkheimer
               L. E. Lewis, Jr.
               Harold T. Newberger
               Elizabeth A. Wood

Publisher: Houghton Mifflin Company
               Boston, Massachusetts

The M.A.P.S. program consists of modules in textbook form. It is a textbook program in that a textbook is used to present the learning experiences, but it isn't a textbook program in that the textbook does not follow the traditional format of presenting information to be read and then verified using the experiments provided. A basic presumption of M.A.P.S. is that children learn by doing—by working with real materials and pictures of real materials, not by just reading about concepts and processes.

Various M.A.P.S. materials.

    There are thirty-one modules in the program. A kindergarten module, *"Learning to Learn,"* starts the program, followed by four modules for each of grades 1-6. Grades 7 and 8 have three modules each. A teacher's guide is available for every level. Each module is based on one of these four concepts:

1. Structural patterns
2. Patterns of space and time
3. Patterns of interaction
4. Patterns of life

Four processes are stressed in each module:

1. Observing and describing
2. Investigating and manipulating
3. Organizing and quantifying
4. Generalizing

The goal of M.A.P.S. is to help students develop their own thinking processes.

Features of the Program

The M.A.P.S. program contains a number of unique features not found in a traditional textbook program. The first, and most prominent, feature is that reading ability is not considered to be important. It is not meant to be a "read-about" program. This brings two important facets into consideration. First, the text provides instructions and commentary, not information, for the students. The instructions and commentary are used by the students to do activities that will help them find the answers to their questions. In effect, information is generated by the students. Second, the students are not inhibited by reading ability. Poor readers, as well as good readers, can *do* science even if they cannot read about it. Part of the teacher's job is to see to it that the students understand the directions. M.A.P.S. does encourage reading and provides aids to insure success, but it does not hinder those students who have difficulty with reading.

Behavioral patterns are also an integral part of the program. Expected student behavior during, as well as at the end of, an activity is given. The behavior patterns are meant to be flexible so that the individual needs of the students can be considered. Evaluation is based on the behavior patterns, and criterion-based tests are available in duplicating master form for formal evaluation of a student's performance.

There is less content in M.A.P.S. than in most traditional programs, but major concepts are explored in greater depth. The feeling is that greater learning will take place if fewer concepts are thoroughly studied.

Another feature of the program is that a variety of teaching techniques are suggested. No one technique is considered to be the best, and the teacher is encouraged to vary the methods used. This also allows the program to fit the teacher rather than trying to make the teacher fit the program.

Illustrations are purposeful, not decorative. Data are presented in the photographs for the students to organize and use in forming conclusions.

A scope and sequence chart in the teacher's manual provides the title of the modules, states the concepts presented in the modules, and describes behaviorally the specific objective of the processes. The teacher can consult this chart and know exactly what is covered in each module and what the students are supposed to learn.

All of these features combine to make the M.A.P.S. program different from the traditional textbook programs and other text-kit approaches as well.

Organization for Teaching

The teacher's guide provides specific information for the teacher. Eight topics are listed in each guide, although not necessarily in the following order. Order is determined by need within the unit.

1. *Overview:* gives the purpose of the module
2. *Background information:* helps the teacher guide the students, is not additional facts for the students
3. *Advance preparation:* gives information about preparations that need to be made for future activities
4. *Concept-process chart:* indicates which specific processes are used to teach the major concepts of the module
5. *List of materials:* identifies the materials needed for the module
6. *Behavioral patterns:* gives a list of the specific behavioral patterns to be achieved in the module
7. *Teaching suggestions:* gives specific ideas for classroom organization, time allowances, key questions, and equipment handling
8. *Feedback opportunities:* suggestions for review of past experiences as well as for preview of coming experiences to help keep the conceptual thread of the module in view

Materials

When discussing materials for the M.A.P.S. program, three different types of materials must be considered. The first is *equipment and supplies.* These are the materials used by the students and teacher in the activities. Two types of kits are available from the publisher:

1. *School kits:* kits containing frequently used materials, common to modules of all levels. There are three kits available: the lens kit, the container kit, and the measurement kit.
2. *Modular kits:* coordinated kits for each level, containing everything you need (except common classroom supplies) to teach the modules. A list is given for each level for teachers who would rather gather their own materials than buy the commercial kits.

The second type of material is the *printed material* available from the publisher:

1. *Consumable books:* paperback books for each module which allow the students to do the activities, then record their answers
2. *Hardbound books:* books containing the four modules for each level. They contain modified directions for recording answers on separate sheets of paper or in the activity and record books.
3. *Activity and record books:* consumables which go with the hardbound books. They provide the answer sheets found in the hardbound books.
4. *Teacher's annotated edition:* provides the teacher's guide which has information needed to teach the modules
5. *Tests:* duplicating masters for coordinated tests

The third type of material available is *audiovisual aids*. Filmstrips and coordinated overhead transparencies are available for levels K-2. A media package for each module is available for levels 3-6, containing filmstrips, sound cassettes, duplicating masters for response sheets, and a teacher's guide. These packages give reading help, tutorial help, enrichment, and additional science background. They can be used individually, in small groups, or by the entire class.

The various materials available are the backbone of the program, but a variety of options allows the school system to put together the type of program it prefers.

You are now ready to do the activities for the M.A.P.S. program, providing you have read the background information. The first activity will be to review one of the teacher's editions. This will help you see what information is given the teacher. Sample activities from the program will follow to let you get a feel for the program from the student's point of view.

Activity 1: Review of M.A.P.S. Teacher's Edition

A. Why should you review one of the teacher's editions of the M.A.P.S. program? Think about the following reasons as you complete this activity:

1. To familiarize you with the M.A.P.S. program and the way it is taught
2. To familiarize you with the type of equipment needed to teach the M.A.P.S. program and how it may be obtained
3. To help you discover a new way to teach elementary science, using hands-on activities and simple equipment

Check with your group and/or your instructor to decide which level you want to work with, then obtain the appropriate teacher's edition.

B. Note the color and grade level of the teacher's edition you are reviewing. Look especially for the following information:

1. What modules are covered in this level?
2. Find the section entitled "Key Features of the Module" and read it. Look for these features as you review the book. Can you find them? Are they really key features from your point of view?
3. Read the section "Getting Acquainted with the Program." Can you find the philosophy of the program? What is the difference between behavioral patterns and behavioral objectives?
4. Examine several lessons in the textbook. What types of materials are necessary for you to teach a lesson? Could you obtain this material very easily?
5. Think about the types of activities presented. Can you devise similar activities?

C. Make some comments about your review. What did you like best about M.A.P.S.? Least? Would you like to use it? These notes should remind you of the program when you refer to them at a later time.

D. Have a group discussion about the M.A.P.S. program. You might want to invite your instructor to participate. Share the comments that you made in C. Make notes on your discussion.

COMMENTS

SELF-CHECK √

Did you read the front material in the teacher's edition or just glance at it? This is the best place to find out what the authors think that the teacher using the program should know about it. Is there anything you would like to know about the program that is not included? (The question of cost can be answered by a sales representative). Did you try one of the lessons? You should do so to learn how the lesson is taught. You will do an activity shortly which is based on activities from the textbook, but it is not the same as actually going through one of the complete lessons in the textbook.

COMMENTS

SUMMARY

You were asked to review a teacher's guide from the M.A.P.S. program to give you some information about it. Your review should have provided you with the author's concept of the program, teaching information, and a view of the children's activities. Discussions with your group and/or your instructor should clear up any misconceptions about the program.

These students are performing the M.A.P.S. activity, Balancing.

Activity 2: Activities for M.A.P.S.

Two M.A.P.S. activities are presented for your hands-on involvement. Your instructor will tell you what is required. Be sure to do one of them or do both of them if time permits. Remember, you are trying to prepare for the day you start teaching, not just fulfilling requirements and trying to get by as easily as possible. Work with your group and share your thoughts. You might want to let one member of your group act as the teacher while the other members go through the activities. Consult with your instructor to see how this might be done.

As you go through this section, think about how you could use these activities, or similar ones, in your classroom. Could you adapt some of these lessons to a special situation such as an inner city school, a rural school, an exceptional class, or one or more unique individuals in the class?

The two activities you will do were selected because of special characteristics and are reprinted directly from the textbook without modification. The first one, *"The Balance Point,"* represents an activity that uses a minimum of equipment to provide a hands-on learning experience. *"Dial-A-Point,"* the second activity, provides you with the opportunity to make a fun type apparatus that you can take with you. It, too, requires few materials, but it provides an excellent learning experience.

The Balance Point

Children enjoy balancing things, and in this activity they are introduced to some of the properties of a balanced object.[2] The purpose of this activity is to introduce the *balance point* of an object. You will need a meterstick or yardstick, several similar coins, and some tape. Obtain these items and begin.

A. The balance point of an object is the point where the object can be supported most easily. Try to balance a meterstick. Place the ends of the meterstick on your two fingers as shown. Close your eyes and slide your fingers toward each other. What happens? Place the ends of the stick on two fingers again. This time slide *one* finger only. What happens? Try the stunt

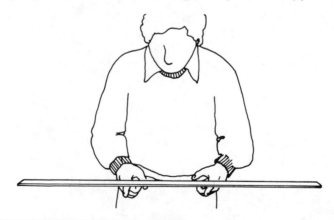

[2]From *M.A.P.S. Teacher's Edition-Level 4* by Carl F. Berger, Glen D. Berkheimer, L.E. Lewis, Jr., Harold T. Newberger, and Elizabeth A. Wood. Copyright © 1974 by Houghton Mifflin Co. Reprinted by permission. P. T-41.

again, using two pencils in place of your fingers. Explain what happens when you use pencils.

Try to balance the meterstick on one pencil. Where is the balance point?

Tape a coin to one end of the meterstick. How will this change the balance point of the stick?

Where is the new balance point?

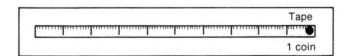

Is it nearer or farther from the coin than the original balance point?

Now tape a second coin of the same kind on top of the first.

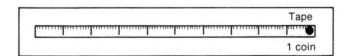

Where is the balance point?

Where do you think the balance point will be with three coins at the end of the stick?

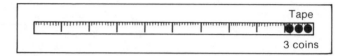

Test your prediction. Where is the balance point?

How could you restore the original balance point of the stick without removing the coins?

√ SELF-CHECK

For the activity:

Did you get the meterstick to balance? You should be able to do it easily enough if it is lying flat but can you do it if it is on edge? You should be able to check your results as each question or instruction leads you through the activity step by step. What do you know about a balance point now?

Implications for teaching:

How could you use this activity in your classroom? Could you expand on it for your faster students? Adjust it for slower students? How many students would you have working together on this activity? Would it help to blindfold the person who is balancing the meterstick?

COMMENTS

SUMMARY

Simple equipment was used to teach the concept of a balance point and some of its characteristics. The step-by-step learning experience provided you with the opportunity to learn about balance points without telling you the answer.

Dial-A-Point

This activity will provide you with an interesting device for teaching polar coordinates, which you can make and take with you.[3] Technically, it can be called a position finder. As you do this activity, you will discover that, unlike the first activity, it is not designed to teach a concept but to provide equipment for interesting puzzles that will lead to future investigations. You will need the following materials:

1. Cardboard strip pointer, 1 cm x 7 cm
2. Corrugated cardboard, 20 cm x 20 cm
3. Masking tape
4. Paper, 20 cm x 20 cm
5. Paper clip(s)
6. Scissors
7. Thumbtack or paper fastener (brad)

Obtain these materials from your instructor and start the activity.

A. There are many kinds of position finders. Here is one you can make. It is called Dial-A-Point.

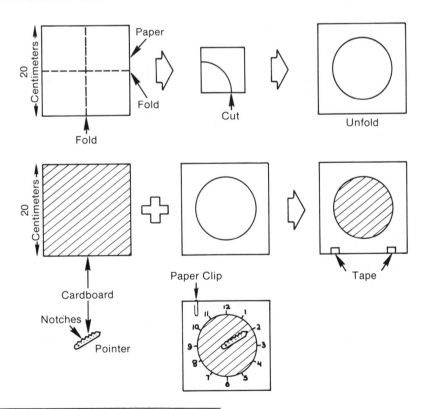

[3]From *M.A.P.S. Teacher's Edition—Level 2* by Carl F. Berger, Glen D. Berkheimer, L.E. Lewis, Jr., Harold T. Newberger, and Elizabeth A. Wood. Copyright© 1974 by Houghton Mifflin Co. Printed by permission. P. T-187.

B. Try a picture puzzle. Use your Dial-A-Point and the chart.

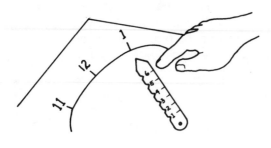

Dial	Point
12	3
1	6
2	3
3	6
4	3
5	6
6	3
7	6
8	3
9	6
10	3
11	6

Dial each number and mark each point.

Remove the pointer and the top.

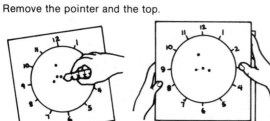

Then connect all the points.
Use straight lines.
Color your picture.
What shape did you make?

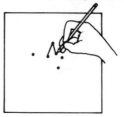

C. Try another puzzle with your Dial-A-Point. Insert a sheet of paper and start.

Dial	Point
11	6
12	6
1	6
2	6
2	6
1	3
12	2
11	3
10	5
10	6

Connect the Dots.

Then Start again with

next column.

Dial	Point
12	2
6	2
6	5
7	4

What did you draw?

D. Make up a puzzle of your own. Have your partners try it. Draw a picture to go with your coordinates.

E. Make an alphabet code using a point for each letter; then try a few secret messages.

SELF-CHECK √

For the activity:

Does your Dial-A-Point work? If not check with your instructor. Did you get an umbrella in part B? Could you give coordinates for a better one? Did you have any trouble making up your own puzzle?

Implications for teaching:

How much fun could your students have with this device in your classroom? It is useful for teaching *coordinates,* which are the reference numbers used to indicate or locate a specific point—usually on a map. Here, you are using polar coordinates of one type. Latitude and longitude are another type of polar coordinates. Could you use the alphabet code to send messages in your classroom?

COMMENTS

SUMMARY

A hands-on device was constructed in this activity, and suggestions were made for its use. A simple way to introduce a complex subject is offered to the novice as well as to the experienced teacher.

SUMMARY: M.A.P.S.

This completed the M.A.P.S. portion of Part 2-3. The background information and the review of one of the teacher's editions of the M.A.P.S. program provided you with the basic information you would need if you were going to teach using M.A.P.S.

The activities gave you the opportunity to make a teaching device for your own use while you learned about position finders and showed you how to use simple equipment to teach some of the concepts of balancing.

The behavioral objectives for this section were that in completing it you will do the following:

1. Identify and describe the materials used by Modular Activities Program in Science (M.A.P.S.)
2. Identify and describe the scope, sequence, and teaching strategy of M.A.P.S.

If you feel that you have reached these objectives, meet with your instructor for a final seminar. If not, review the sections that are not clear to you. Ask your instructor for help if needed.

FINAL SEMINAR

A review of the background information and activities should refresh your memory about the M.A.P.S. program and prepare you for the final seminar. Meet with a small group of classmates to discuss your impressions of this program. *Don't forget to invite your instructor to participate in this final seminar.*

NOTES

BIBLIOGRAPHY

Berger, Carl et al., *Modular Activities Program in Science (Teacher's Annotated Edition).*
    Boston, Mass. Houghton Mifflin Co., 1974.

SUGGESTED READING

Berger, Carl et al. *Modular Activities Program in Science (Teacher's Annotated Edition).*
    Boston, Mass.: Houghton Mifflin Co., 1974.

Summary: Text-Kit Approach to Elementary Science

In this Part, you were introduced to two programs that were a combination of a textbook program and a kit program. Each of them tried to combine the best features of both approaches to make science meaningful to both the teacher and the student. Hands-on manipulation of common materials provides the opportunity for learning science processes and gives the concrete experiences considered vital by Piaget and others. The textbook, on the other hand, provides the structure, direction, and stability that many teachers and students need to help them function.

The *Elementary School Science: Space, Time, Energy, and Matter* (STEM) program of Addison-Wesley Publishing Company uses four concepts—space, time, energy, and matter—as its basis. It stresses equal development of the intellectual and process skills. Reading materials in the text provide the intellectual material; hands-on activities provide the process skill development.

The Modular Activities Program in Science (M.A.P.S.) by Houghton Mifflin Company uses the textbook to present the learning situation to the students so that they can learn, not by reading about a concept, but by actually getting involved in inventing the concept. The text provides the skeletal structure upon which the students build understanding. Words and pictures provide the necessary input to get the students started in the right direction so that they can formulate their own concepts.

The background information you were given at the beginning of each section provided you with material about each of the programs. The activities gave you the opportunity to review the teacher's editions of each program to further your knowledge about the programs, and the hands-on portion let you try some of the same activities the students would use.

If you were satisfied, upon completion of each section, that you had achieved the indicated objectives, then you should be able to say that you have achieved competency in this Part. If you are ready, arrange for a final seminar with your instructor.

FINAL SEMINAR

Review your notes from your discussions of the STEM and M.A.P.S. programs to prepare yourself for the final seminar. This seminar may be either a small group review, or the instructor may lead a large group review. *Check with your instructor for directions.*

NOTES

COMPETENCY EVALUATION

The background information and activities for this Part were designed to familiarize you with text-kit materials that can be used to conduct learning experiences with elementary school students. Your instructor may choose to administer a competency evaluation measure of some type. Consult with him for specific directions.

UNIT 3
Program Management

It is necessary that you become skillful in managing your science program. No science program or textbook, no matter how well written, can be expected to fit the individual needs of all children in a specific classroom. The teacher must be able to make the necessary adjustments and modifications so that optimum utilization is possible. Skillful management of the science program can mean the difference between effective, successful science experiences and mediocre ones.

The first step in the skillful management of your science program involves becoming knowledgeable about the various elements of elementary science instruction. Units 1 and 2 provided you with many opportunities to become directly involved with these various elements. In Unit 1, you were involved in activities designed to help you achieve competence in the area of sciencing. The specific data-gathering and data-processing skills needed for achieving competence in this area were identified. You may want to refer to pages 3-4 to refresh your memory. These process-inquiry skills are the tools of scientific investigation, which are used to gather, organize, analyze, and evaluate science content.

In Unit 2, you were involved in examining science curricula materials. Three types of materials were presented: laboratory or kit materials, text-books, and text-kit materials. The purpose of that Unit was to familiarize you with the scope and sequence of these programs as well as with the teaching strategy and materials used in each.

The second step in the skillful management of your science program involves the application of the information and ideas that you were exposed to in the first two Units. Unit 3 is designed to help you accomplish this step.

PART 3-1

Managing Your Science Program

FLOWCHART: Managing Your Science Program

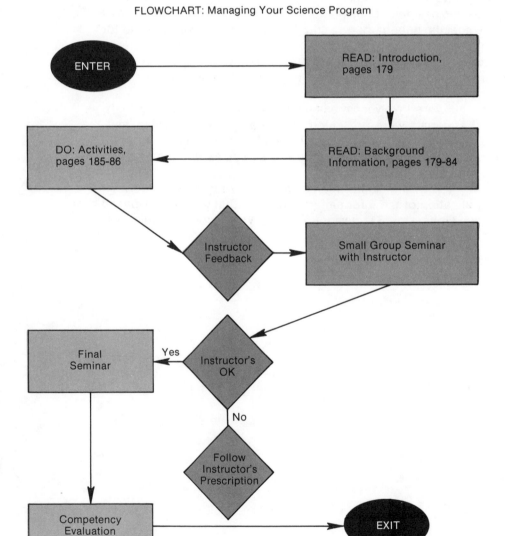

Introduction

The information and ideas that you obtained from Units 1 and 2 can be applied in three ways. First, you can use the information to help you get the most out of an existing science program. Second, you may be involved in selecting a new science program for your school. Consequently, your knowledge can help you make an intelligent contribution. And third, if you have to develop your own science program where none exists, you should be able to do so. This Part will explore these three possibilities.

GOAL

After completing this Part, you will demonstrate competency in the ability to utilize acquired information, ideas, and skills in developing science program management skills.

BEHAVIORAL OBJECTIVES

In completing this Part, you will do the following:

A. Identify and explain how an existing science program can be modified to meet classroom needs

B. Identify and describe how acquired ideas, information, and skills can be used in selecting an appropriate science program for a specific school system

C. Describe how acquired ideas, information, and skills can be used in helping you develop your own science program

---

Background Information

Using an Existing Program

Most schools have a science program of some type. If you are fortunate enough to teach in a school that has a good, strong science program based on one of the new science programs or one of the current textbook programs, you can follow the teacher's guide and do a reasonable job of teaching. But what if you are given an obsolete or, in your opinion, a very poor or weak textbook? Several options are open to you. First, you can panic. This does no good at all and is not recommended. Second, you can stack the books in the corner and conveniently forget to teach science. This, too, is a poor choice and is not recommended. The third choice is to go ahead and follow the textbook, like it or not. Although not the best choice, it is better than the first two options because your students will get some science instruction. The fourth option is to use what you have to help you develop your own science program. This is what this section is all about.

What can you do with what you have? As a beginning teacher, you should talk to other teachers or the principal to find out how you are expected to use the textbook. They can tell you whether you are supposed to follow it closely, if you can digress somewhat, or if you are free to develop your own program. This is important because you do have to meet the expectations of the school system that employs you. Do remember though that you can exceed the expectations by supplementing the minimum requirements. Develop a strong science program with what you have or can "buy, beg, or borrow."

Here are some suggestions to help you with your science program. Use any of them that are appropriate to your situation. They can apply to *any* textbook series, good or bad, that you may have in your classroom.

1. Use the textbook to help you decide what to teach. In most school systems, the adopted textbook series acts as a curriculum guide, establishing the philosophy, scope, and sequence of the science program. Your textbook can tell you what content you are expected to cover. Look at the table of contents to see what topics are listed and use this as your guide. The actual material in the text can help you decide how thoroughly, or superficially, you should treat each topic.
2. Use your textbook as a resource. Even the poorest textbook can serve in this capacity. It can provide you with teaching suggestions, experiments and demonstrations, suggested reading lists, teaching aids, and background information. You don't have to use every suggestion or activity. Be selective, but don't ignore this source just because it is old or you don't like the color of the cover.
3. Develop some hands-on experiences for your students. If activities in your textbook are nonexistent or teacher oriented, use what you have learned here to develop new, or modify existing, activities to involve the students. Look for ways to involve them in concrete experiences rather than the more traditional abstractions. Let the students do the demonstrations and experiments rather than doing them yourself.
4. Find a way to include the process skills in your science program. Some textbooks include these skills, but many do not. Process skills are the tools of the scientist, and children should learn to use them automatically when presented with a problem or a learning situation.
5. Use trade books to supplement your science program. (Trade books are any book other than textbooks.) They provide a source of factual information, experiments, and reading activities for you and your students. The school library should have a good selection of these books. In addition, a classroom library might be developed from books brought from home by the students.
6. Bring in outside speakers. Use parents, as well as other members of the community, to present information to your students. You have access to scientists, professional people, and knowledgeable amateurs, if you will look around the community for them. Let them help you teach science by bringing in expertise that you and your textbook may not be able to provide.
7. Try to provide something for everyone in your classroom. Most textbooks are written for the average student. You must have activities simple enough for your slower students as well as enrichment activities for your gifted students. Every child should be able to work at a level commensurate with his ability. Students

who cannot do the work assigned or who find no challenge to it are not being treated fairly by the teachers.

You may want to add to this list of suggestions. Remember, whatever happens in your classroom is directly related to what you plan and do there.

Selecting a Science Program

If you are employed by a school system that does not have a science program or wants to replace an existing program, you may be asked to serve on a science selection committee, or at least to give your opinion to the committee. This is your chance to contribute to the improvement of science education. Obviously, you want to recommend only the best, but which is best? Although it is often a temptation to ask your college instructors or others for an authoritative opinion, you must decide for yourself what you mean by best. Best for whom? Then you can decide upon the program that "best" fits the needs of the user.

All of the programs and textbooks are good, or they would not be favorably received on the national market. There are many school systems using each with great success. The trick is to fit the program to the school. When you start selecting a new program, it may be easy to decide which program you like best. But, more important, you must decide which one is most appropriate for the kind of students you teach. The preferences of other teachers in the school also must be considered. A program must be selected that they can, and will, use. Remember, no program is effective if it is not used.

Several lists of criteria are available to help you select an appropriate program or you may even want to develop your own list. Here are two lists that might be of help when you consider the various programs for adoption.

The first one was developed and used by the Mesa County Valley School District No. 51 in Grand Junction, Colorado, and was reported in *Science and Children*. A science committee was established and given the task of developing criteria and reviewing science programs. Here are their criteria.

Selection Criteria

The primary goal of the science committee was to recommend a science program that could be taught by educators representing a variety of teaching philosophies. To meet this goal, and to give each program under consideration an equal evaluation, twelve basic criteria were established. A comparison of all elementary school science programs evaluated was possible by recording the results in chart form.

1. *Level of Readability.* A majority of the teachers surveyed felt that reading and science were closely interrelated. The teachers requested a science textbook that was written at a grade level that was consistent with their respective reading programs. The general consensus was that reading would be reinforced.

2. *Durability of the Program.* As a cost-saving factor, the committee studied each series in terms of its physical construction. This analysis went beyond the textbook analyzed, and was applied to all instructional material within the science program. Some specifics researched were as follows: Was the text a paper or hardbound design? If a laboratory kit was included, was it sturdy, or would replacement of parts be an ongoing cost? How was the teachers' manual bound? (paper back, spiral, or hardbound?)

3. *Preparation Time.* The teachers surveyed felt that science rated well behind mathematics, reading, language arts, and other instructional areas in terms of their priorities and time expended. The preparation time devoted to science instruction was to be minimal. Any materials to be supplied for use in an experiment should be easily obtained.

4. *Text Format.* The committees desired to afford teachers a degree of flexibility in the utilization of the text. A textual program which contained the unit format was desirable. The order in which the units were to be taught would be left to the individual teacher. If district-wide uniformity were needed, a specific order for unit presentation could be established. This would help in cases of in-district transfer of children.

5. *Teacher's Manual Format.* Because preparation time devoted to science is normally minimal, the committee searched for a comprehensive teacher's manual. Were teaching suggestions included on the student pages? Were materials needed for each lesson listed? Were goals and possible outcomes of each lesson provided? Were lessons clearly outlined?

6. *Laboratory Kit.* If a program included a laboratory kit, could the kit be utilized by more than one classroom at a time? Was the kit set up so that students could maintain and inventory it? Were the components of the laboratory sturdy enough to stand up for an extended period of time? The committee felt that a kit which could be easily maintained, was durable, and would require little replacement of component parts would be readily accepted.

7. *Testing Materials.* Was a set of testing materials included with the program? Although many teacher-made evaluation materials were good, a comprehensive evaluation system would insure more valid testing results. To obtain some degree of grading uniformity across the district elementary schools, a comprehensive set of test materials should be included with the basic instructional program.

8. *Kindergarten Materials.* Although many elementary science programs do not include the kindergarten level, the committee deemed this to be desirable. The kindergarten readiness for science should be conducted in the same program that would be used in the later grades. Within a single program, major concepts could be introduced in kindergarten. Some degree of program uniformity in science content and how it is presented could thus be established from the very beginning.

9. *Supplementary Books.* Are any books other than the textbooks required. If a laboratory book or a record book is included, is it necessary to the program? If another book is needed, is it reusable or consumable? Audiovisual and manipulatory materials for the program could be subjected to this test.

10. *Health.* Since science and health are frequently scheduled into the same block of time for instructional purposes, the committee sought a science program which included some large amount of health education. If money could be saved by not purchasing a separate health series, more funds would be available for the science materials.

11. *Content and "Hands-On."* Since the district employs teachers who use a wide variety of instructional approaches, how can one science program meet their instructional needs? The most acceptable answer was the identification of a science program which was well balanced between content or printed information, and "Hands-On" or experiment oriented material.

12. *Experiments.* Because the district wanted a program which included experiments as a part of the learning process, there was the need to evaluate the kind of experiments which were included. The committee looked for purposeful experiments which require some limited amount of preparation time. The materials required should either be included in the laboratory kit, or easily and inexpensively obtained. The

committee desired open-ended exploration, rather than a guided step-by-step approach.[1]

The second list was presented in *The Science Teacher*. It was developed for use by teachers as a guideline for evaluating science programs. Although oriented more toward textbook selection than toward the laboratory kits or teacher-developed science programs, with a little modification, the list of criteria can apply to any science program.

What to Look for in Science Programs

1. What is the underlying philosophy of the program, and do you feel comfortable with it?
2. Do the materials communicate their intent to the student, and do the teacher materials communicate their message to the teacher?
3. Is the program practical in terms of its aims, teachability, application to student's everyday living, and in building a foundation of science knowledge?
4. Does the program permit flexibility in teaching styles, sequence, time, content, classroom facilities?
5. Does the program appear to be appealing to the students?
6. Is there a variety of science topics and are the topics appropriate?
7. Does the program include opportunities to apply scientific methods?
8. How are the various parts of the program organized, and do you feel comfortable with the arrangements?
9. Are tests and other assessments provided?
10. Are learning objectives for the program stated, and do you feel they are appropriate and reasonable?
11. Is the student reading material at a reasonable level, and is there a good balance between "reading" and "doing" activities?
12. To what extent are students directly involved with "things" and laboratory activities, and do you consider it sufficient?
13. Are there provisions for individual student differences?
14. Are the time requirements to teach the material on a day-to-day or week-to-week basis reasonable?
15. Does the teacher's guide provide helpful suggestions relating to teaching strategies, student activities, methods of evaluation, use of audiovisual aids, and other material?
16. Does the program provide for the needed mathematics?
17. Does the program give you a feeling that you could successfully teach science?[2]

Criteria lists are a helpful guide, but you also have to consider which of the criteria you wish to place the most, and which the least emphasis on. You and the other members of your faculty can use the same list, and the result will be entirely different opinions of the material being evaluated. A good selection can only be made after much evaluation, discussion, and give-and-take among all of those teachers who are going to use the program.

---

[1] Wayne Reeder and Ben Adams, "Selection of an Elementary Science Program: Process and Criteria," *Science and Children* 14 (March 1977): 8-10.

[2] James V. DeRose and Don P. Whittle, "Selecting Textbooks: A Plan that Worked," *The Science Teacher* 43, No. 46 (September 1976): 40.

One final point, try to look at as many programs as you can before making a selection. Too many schools limit themselves to the consideration of only a few textbooks from some of the more familiar publishers. They may not be aware that other options exist. One of the purposes of Unit 2 was to make you aware of the alternatives available so that you can make intelligent choices.

Developing Your Own Science Program

Not all schools have an established science curriculum. There may be various reasons for the lack of an established program. Some schools may furnish teachers with many resources and basic curriculum guidelines. Then, the teachers are expected to use these materials in developing an innovative science program.

Other schools may provide teachers with a good basic program and give the freedom to use it totally, in part, or not at all in providing science experiences for the students. In a school where a weak or outdated program exists, the teacher has the responsibility of developing supplementary experiences. In a few cases, there may be *no* program at all, and the teacher must start from scratch, In any case, you will face a challenging, but rewarding, opportunity.

In developing your own science program, you will need all the help you can get. There are several good sources. In all probability, you will go to a textbook first. This is an excellent place to start because it can provide you with a course outline, teaching suggestions, experiments and demonstrations, and factual information. Your fellow teachers are also an excellent source of help. There is nothing wrong with looking to more experienced teachers for information and helpful suggestions. It is only when you depend on others for everything that you are in trouble. Teachers who have been in the school system for a few years can help you find equipment, tell you what units should be taught, and offer teaching suggestions. If consultants or resource people are available, don't hesitate to use their services. Too often, beginning, and even experienced, teachers are reluctant to ask for help because they are afraid it will be construed as a sign of weakness or inability to do the job they were hired to do. This is nonsense. Good teachers use every tool, idea, or resource available to help improve their teaching. A third, and infrequently used, resource is the children in your classroom. Find out what their interests are and what problem-solving skills they have. Look for hidden talents that might be useful. Above all, don't underestimate the contributions that your children can make or what they can teach you. Good teachers learn as they teach. The final source is, of course, you. You will have to draw upon what you have learned in your teacher preparation courses, student teaching, and personal experiences to help you assimilate and use the information you have gathered.

Activity 1: Developing Activities

A. Working alone, select two science concepts from those listed below. Choose one from the physical science area and one from the life science area. Develop an activity for each that would require elementary children to use at least two of the process-inquiry skills in gathering, organizing, analyzing, or evaluating the science content. You may use any of the curricula materials that you examined in Unit 2 in developing these two activities.

Physical Science Concepts

1. Air takes up space and has weight.

2. When an object vibrates, sound may be produced.

3. Simple machines help in moving objects.

4. Light appears to travel in a straight line.

5. Soil consists of several different layers.

Life Science Concepts

1. Insects change in body form until they become adults.

2. Plants have leaves, roots, stems, and flowers.

3. Mold needs warmth, moisture, and usually darkness in order to grow well.

4. Different environments are needed to sustain different types of life.

5. Some plants grow from seeds.

B. Share your activities with a group of your classmates. Ask them to give you some feedback. Make any changes that you think would improve your activities based on the feedback from your group.

Activity 2: Implementing Activities

A. After you receive evaluation and feedback from the instructor on your planned activities, ask about opportunities for implementing at least one of the activities. Many possibilities exist. If you know and have access to elementary children, you might arrange to carry out your planned activity outside of class. There may be local schools close by which would allow you to work with a small number of children during regular school hours. Another possibility is inviting an entire class from a local school to visit your science class during a regularly scheduled class period. The children could form small groups and participate in the different activities set up and conducted by you and your classmates in different parts of the classroom. You will probably think of many other ideas to implement the activities you have planned. Discuss your ideas with your instructor and be prepared to implement your activity at the time and place agreed upon.

B. Discuss the outcome of your planned activities with a small group of your classmates. Invite your instructor to share in your discussion, too. Use the following questions to help structure your discussion:

1. Was the science experience a successful one for you?
2. Was it successful for the children?
3. What would you do differently, if you did it again?
4. Were you able to incorporate at least two of the process-inquiry skills into the experience?

C. Discuss the information and ideas presented in these first three Units and suggest ways to utilize them in managing your science program.

COMMENTS

Summary: Managing Your Science Program

Skillful management of your science program is a necessary component for meaningful science experiences. As was stated earlier, no science program or textbook, no matter how well written, can be expected to fit the individual needs of all children in any specific classroom. This means that you, the teacher, must have the skill to make adjustments and modifications in existing programs so that optimum utilization is possible.

This Part provided you with additional information and ideas to help you apply the information and skills that you acquired in the previous Units. Three suggestions were given. First, the information can be used to help you get the most of an existing science program. Second, it may be useful in helping you select a new science program. And third, you may use it in developing your own science program where none exists.

FINAL SEMINAR

Review this Unit and look over the questions and comments you recorded after the various activities. If you have specific questions, share them with a small group. Together you might be able to generate some answers. Invite your instructor to join you for this final seminar.

NOTES

COMPETENCY EVALUATION

Your instructor may choose to use a competency evaluation measure of some type to evaluate your competency in the area of managing your science program. Check with your instructor for specific directions.

BIBLIOGRAPHY

DeRose, James V, and Whittle, Don P. "Selecting Textbooks: A Plan that Worked." *The Science Teacher* 43, no. 46 (September 1976): 40.

Reader, Wayne, and Adams, Ben. "Selection of an Elementary Science Program: Process and Criteria." *Science and Children* 14, no. 8 (March 1977): 8-10.

UNIT 4
Classroom Methodology

This Unit will provide you with some current ideas and learning theories pertaining to methods of classroom instruction. Part 4-1 is designed to help you construct effective daily teaching strategies. Three stages of lesson planning are examined:

1. Developing and writing objectives
2. Selecting and developing appropriate teaching techniques
3. Utilizing evaluation techniques

Each stage is discussed, and activities are provided to engage you in a step-by-step process of lesson planning. The end result is a teaching strategy for a science activity that includes instructional objectives, behavioral indicators, techniques or methods to be utilized, and evaluation to be employed.

Part 4-2 focuses on questioning techniques. Various questioning strategies that offer guidance to elementary teachers in planning and integrating effective questions into the instructional strategy are examined. Guidance in responding to students' replies to higher-level questions is also included. The activities involve you in formulating questions that foster creative, thoughtful, high-level responses on the part of elementary students.

PART 4-1

Planning for Teaching

FLOWCHART: Planning for Teaching

```
  ENTER  ───────────────────────►  READ: Introduction,
                                    pages 191
                                         │
                                         ▼
                                    READ: Background
                                    Information "Developing and
                                    Writing Objectives," pages 192-99
                                    DO: Activities, pages 200-02
                                         │
        Check with                       ▼
        Instructor for  ◄───────    READ: Background Information
        Feedback                    "Selecting and Developing Appropriate
                                    Teaching Techniques," pages 204-12
                                    DO: Activities, pages 213-15
                                         │
                                         ▼
                                    READ: Background Information
                                    "Utilizing Evaluation Techniques,"
                                    pages 217-20
                                    DO: Activities, pages 221-23

  Small Group          Instructor's    Yes      Competency
  Seminar with    ───► OK          ──────────►  Evaluation
  Instructor                                        │
                          │ No                       ▼
                          ▼                        EXIT
                       Follow
                       Instructor's
                       Prescription
```

Introduction

The most effective and successful learning experiences take place as a result of systematic, creative planning. There are three stages of lesson planning: (1) developing and writing objectives, (2) selecting and developing appropriate teaching techniques, and (3) utilizing evaluation techniques. Consequently, this Part is divided into three sections in order to focus on each of the three stages of lesson planning.

The first section is concerned with developing and writing objectives. Objectives serve to identify the purpose of instruction. In order to provide elementary children with meaningful science experiences, a teacher needs to identify the purpose of the experience. What is supposed to be accomplished? How will the child benefit from the experience? Objectives should provide answers to these questions.

The second section continues the planning process by offering guidance in selecting and developing teaching techniques. After establishing the objectives of the lesson, the teacher must devise an instructional sequence that will aid the learner in attaining the objectives. A variety of teaching techniques that can be utilized in planning for the instructional portion of the lesson are presented in this section.

Finally, the third section provides insight into the utilization of evaluation techniques. One of the most difficult tasks a teacher faces is that of assessing the learner's achievement; yet, it is an essential part of all science instruction. Evaluation assumes two functions. One is to provide feedback *during* instruction, and the other is to provide a basis for assigning a grade *after* instruction. A variety of evaluation techniques are presented in this section.

Hopefully, after completing this Part, you will understand and utilize current ideas and learning theories in developing and constructing effective daily teaching strategies.

GOAL

After completing this Part, you will demonstrate competency in the ability to incorporate the current ideas and learning theories presented into effective daily strategies for involving elementary children in sciencing activities.

BEHAVIORAL OBJECTIVES

In completing this Part, you will do the following:

A. Identify and explain the learning theories and ideas presented
B. Describe the implications that the theories of Piaget and Bruner have for constructing teaching strategies
C. Utilize the ideas presented here by selecting a science topic and constructing a teaching strategy which includes instructional objectives, behavioral indicators, techniques, or methods to be utilized and evaluation to be employed

Developing and Writing Objectives

Background Information

Before you attempt to teach a lesson, ask yourself the following questions:

1. What am I trying to accomplish?
2. How do I want the students to be different at the end of the lesson?

These questions can help you organize your efforts and provide children with meaningful instruction. When teachers just follow the instructions in the teacher's manual, without first asking the above questions, learning can become purposeless and haphazard.

The answers to the above questions identify your goals and objectives. For example, suppose you want to teach your students about mold growth. When you make your plans, you need to identify the purpose of the lesson. To make the purpose even more meaningful, you need to think about it in terms that relate to the learner; e.g., "How will the learner be different after the instruction?" Possible answers might be these:

1. To teach students that there are many different kinds of molds
2. To show students that molds have different colors
3. To help students see that certain factors can retard or speed up mold growth
4. To help students discover the factors that affect mold growth
5. To encourage students to investigate and gather other information about molds

These answers can serve as general objectives for a lesson. They can be used to give direction and purpose to the instruction being developed. Since these objectives relate to the instructional part of the lesson and not directly to the student's behavior, the term *instructional objectives* can be used to describe them.

Although these objectives may be desirable and helpful in developing instruction, they provide very little direction as to how a teacher can determine whether or not the students have met the objectives. In order for these objectives to be more useful, they must project specific outcomes of learning. Stating an objective in terms of observable student behavior not only provides direction for the instructional phase, but also for the appraisal of the effectiveness of the learning experience—the evaluation. *Behavioral objectives* offer a format for stating goals in terms of observable student behavior. A complete behavioral objective contains the following information:

1. Identifies observable behavior students will exhibit to show the objective has been fulfilled
2. Identifies the conditions (when and where) under which students are to exhibit the behavior

3. Describes how well students must perform

Behavioral Objective

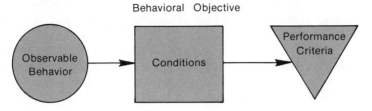

In the activity section, you will develop and write behavioral objectives. The objectives you write here only need to identify the observable behavior the student will exhibit to show the objective has been fulfilled. Criteria 2 and 3 will be fulfilled when you develop evaluation techniques. The behavioral objective should be developed from the instructional objective by asking, "What do I want the students to do when the instruction phase is finished that they couldn't do before?" Remember to state the answer in observable student behavior. Certain terms are useful to denote specific observable behavior. Some of these include the following:

compare	list	differentiate
group	state	compose
create	identify	name
measure	describe	control variables
hypothesize	construct	apply
organize	explain	order
write	distinguish	demonstrate
interpret	perform alone	evaluate

In order to get some practice in writing behavioral objectives, rewrite the proposed instructional objectives for the lesson on molds using the terms listed above so they are stated in observable student behavior. The objectives are listed again for your convenience.

1. To teach students that there are many different kinds of molds.
   Behavioral Objective: The student will

2. To show students that molds have different colors.
   Behavioral Objective: The student will

3. To help students see that certain factors can retard or speed up mold growth.
   Behavioral Objective: The student will

4. To help students discover the factors that affect mold growth.
   Behavioral Objective: The student will

5. To encourage students to investigate and gather other information about molds.
   Behavioral Objective: The student will

You might want to take some time at this point to share your results with your classmates and/or instructor in order to get some immediate feedback. Once you have mastered the task of stating objectives in observable student behavior, you are ready for the next phase of developing and writing behavioral objectives—selecting *significant* objectives.

You should notice that the verbs listed above differ in complexity and cognitive level. Some—*name, state,* and *list*—for example, call for memory level cognition while others—*distinguish, apply,* and *demonstrate*—demand higher levels of cognitive functioning. Just stating objectives so that they describe specific observable learner behavior does not assure that an objective is significant and valuable. In fact, the types of behavior most easily described are often the most trivial. Significant objectives are frequently elusive and difficult to state behaviorally. However, there is help available in determining what is and what is not significant.

Objectives can be divided into four main behavior categories:

1. Cognitive content
2. Cognitive process
3. Psychomotor
4. Affective

Cognitive

Cognitive content objectives identify what you want the students to know. These objectives usually deal in facts and simple concepts that students commit to memory. Such objectives are a necessary part of building a foundation for higher levels of learning. But, you must be careful not to limit your objectives or overemphasize this area. Cognitive content objectives are the simplest and easiest to develop and write but can often be the most trivial.

Cognitive process objectives identify inquiry skills that you want the students to develop and improve as a result of your instruction. These objectives involve the basic processes of observing, classifying, measuring, using spatial relations, communicating, predicting, inferring, and the integrated processes of defining operationally, formulating hypotheses, interpreting data, controlling variables, and experimenting. Cognitive process objectives are more difficult to develop and write than are cognitive content objectives and usually involve a higher level of cognition. In this behavioral category, students must do more than recall some information; generally, they are involved in collecting, organizing, analyzing, and evaluating the information.

The *Taxonomy of Educational Objectives, Handbook 1: Cognitive Domain* provides the teacher with "a range of possible educational goals or outcomes in the cognitive area." A brief modified outline of this taxonomy is included here to provide you with criteria to use in selecting and developing both cognitive content and cognitive process objectives.[1]

---

[1]From *TAXONOMY OF EDUCATIONAL OBJECTIVES; The Classification of Educational Goals: HANDBOOK I: Cognitive Domain,* by Benjamin S. Bloom et al. Copyright © 1956 by Longman Inc. Reprinted by permission of Longman.

I. *Knowledge* (recall of information)

   A. Knowledge of specifics (isolated bits of information)

      1. Knowledge of terminology

      2. Knowledge of specific facts (dates, events, persons, places, etc.)

   B. Knowledge of ways and means of dealing with specifics (passive awareness of how to organize, study, judge, etc.)

      1. Knowledge of conventions (awareness of characteristic ways of treating and presenting ideas and phenomena—correct form and usage in speech and writing)

      2. Knowledge of trends and sequences (awareness of the processes, directions, and movements of phenomena with respect to time)

      3. Knowledge of classification and categories (awareness of fundamental classes, sets, or divisions for a given area)

      4. Knowledge of criteria (awareness of existing criteria by which judgments are made)

      5. Knowledge of methodology (awareness of the skills of inquiry)

   C. Knowledge of the universals and abstractions in a field (awareness of major sciences and patterns by which phenomena and ideas are organized)

      1. Knowledge of principles and generalizations (awareness of the abstractions which summarize observations of phenomena)

      2. Knowledge of theories and structure (awareness of the body of principles and generalizations and their understanding information)

II. *Comprehension* (lowest level of understanding information)

   A. Translation (ability to alter the form of communication without changing the original idea—paraphrasing, putting verbal material into symbolic statements, etc.)

   B. Interpretation (ability to reorder, rearrange, or provide a new view of original material—explaining, summarizing)

   C. Extrapolation (ability to go beyond the original data and determine implications, consequences, effects, etc.)

III. *Application* (ability to use abstractions—general ideas, rules of procedure, etc.—in specific and concrete situations)

IV. *Analysis* (ability to separate a communication into its parts in order to see its organization, effects, basis, and arrangement)

V. *Synthesis* (ability to arrange and combine parts, pieces, or elements into a pattern or structure not clearly there before)

   A. Production of a unique communication (ability to convey ideas, feelings, and/or experiences to others)

   B. Production of a plan or proposed set of operations (ability to develop a plan of operations which satisfies the requirements of the task)

   C. Derivation of a set of abstract relations (ability to develop or deduce a set of abstract relations to explain specific data)

VI. *Evaluation* (ability to make judgments using criteria

   A. Judgments in terms of internal evidence (ability to evaluate a communication using logical accuracy, consistency, and other internal criteria)

   B. Judgments in terms of external criteria (ability to evaluate a communication using selected or remembered criteria)

The six main subdivisions of this taxonomy,—knowledge, comprehension, application, analysis, synthesis, and evaluation—are sequenced so that *knowledge* represents the simplest behavior, and *evaluation,* the most complex. However, the subdivisions are not mutually exclusive; many of the simple behaviors are included in the more complex behaviors. Cognitive content objectives generally fall into the knowledge category of Bloom's *Taxonomy,* while cognitive process objectives are grouped under the other five categories.

Psychomotor

Psychomotor objectives identify manual, manipulative skills that you want the students to develop or improve. Concern with this area of learning seems to be growing. Many early childhood programs focus on the development of movement behaviors. Curricula designed for children with learning difficulties reflect this emphasis as do curricula for vocational education, the fine arts, and physical education. In elementary science, psychomotor objectives identify the skills involved in the proper use of science equipment and materials. Generally, these objectives are not difficult to write since, unlike the cognitive and affective areas, they deal directly with observable behaviors.

Psychomotor behaviors are very important in a process-oriented science approach where the students are involved in gathering data. The students must learn safe and correct manipulation of science equipment and materials in order for maximum learning to take place. A modified psychomotor taxonomic model is included here to help you develop and write psychomotor objectives. This modified model does not include the first two categories of the domain: reflex movements and basic-fundamental movements. Behaviors characteristic of these two categories are innate and, therefore, not considered a part of the learning process.

I. *Perceptual Abilities* (ability to interpret stimuli and respond appropriately)
   A. Kinesthetic discrimination (awareness of the body, how it moves, its position in space and its relationship to the surrounding environment—balance)
   B. Visual Discrimination [ability to (1) receive and differentiate among various observed objects, (2) follow symbols or objects with coordinated eye movements, (3) recall from memory past visual experiences or movement patterns and reproduce them, (4) identify the dominant moving object and respond to it, and (5) be consistent in visual interpretations of the same type of object]
   C. Auditory discrimination [ability to (1) receive and differentiate among various sounds, (2) distinguish the direction of a sound and follow it, and (3) recognize and reproduce past auditory experiences]
   D. Tactile discrimination
      (ability to discriminate among varying textures)
   E. Coordinated abilities
      (ability to coordinate perceptual abilities with movement patterns—eye-hand or eye-foot)
II. *Physical Abilities* (ability to make skilled movements)
   A. Endurance
      (ability to continue activity for long periods of time)
   B. Strength
      (ability to exert maximum force by muscle or muscle group—arm, leg, or abdominal strength)

C. Flexibility
(ability to utilize the range of motion in the joint to produce efficient movement)

D. Agility
(ability to move quickly, change directions, stop and start, reaction—response time, dexterity)

III. *Skilled Movement* (ability to perform with proficiency a complex movement task)

A. Simple adaptive skills
(ability to modify a basic movement pattern to fit a new situation)

B. Compound adaptive skills
(ability to utilize an implement in performing a movement task)

C. Complex adaptive skills
(ability to perform complex movement behaviors which involve total bodily movement, generally without a base of support—gymnastic stunts, twisting dives, etc.)

IV. *Nondiscursive Communication* (ability to communicate through movement behaviors)

A. Expressive movement
(ability to communicate through bodily expressions—body posture and carriage, gestures, and facial expressions)

B. Interpretive movement
(ability to communicate through aesthetic and creative movements)[2]

The four categories included in this modified taxonomy are considered hierarchical in nature. Category I—Perceptual Abilities—represents the lowest level of learned manipulative ability, and Category IV—Nondiscursive Communication—represents the highest. Two of the categories—Physical Abilities and Nondiscursive Communication—are not extensively utilized in science activities but were included so that you could gain insight into the various components of the psychomotor domain.

Most psychomotor science objectives fall into the two categories Perceptual Abilities and Skilled Movements. Perceptual Abilities encompasses those science objectives that relate to the process of observing. The ability to use the senses of sight, hearing, and touch when gathering data is an important science objective. Objectives that concern correct and safe manipulation of science equipment and material fall under Skilled Movements.

Affective

Affective objectives identify attitudes, feelings, emotions, and values that you want the students to develop as a result of your instruction. These objectives are frequently the most difficult to develop and write in behavioral terms. In the past, many teachers have avoided this behavior category altogether. However, leading theorists, such as Carl Rogers, Abraham Maslow, and Art Combs, suggest that affect plays an integral role in determining the total learning outcome.[3] As teachers, then, we must concern

[2]From *A TAXONOMY OF THE PSYCHOMOTOR DOMAIN: A Guide for Developing Behavioral Objectives,* by Anita J. Harrow. Copyright© 1972 by Longman Inc. Reprinted by permission of Longman.

[3]For more information, see books by Rogers, Maslow, and Combs in the Suggested Reading for this section.

ourselves with *all* aspects of the learning process—affective as well as cognitive and psychomotor.

The problem becomes one of identifying appropriate feelings, emotions, values, and attitudes that you feel are important for children to develop as a result of your science instruction and then of putting them in terms of observable student behavior. Because of the very nature of attitudes, values, feelings, and emotions, we can only infer their presence through outward behavior. Therefore, in developing and writing affective objectives, you must determine what behaviors you are willing to take as evidence that a student possesses the attitude, value, feeling, or emotion you have selected.

David Krathwohl and Benjamin Bloom, both of whom were involved in producing the original taxonomy for cognitive objectives, were joined by Bertram Masia in preparing a second handbook. *Handbook II: Affective. Domain* provides teachers with an organized structure to use in defining affective objectives. It also offers help in identifying techniques for evaluating affective behavior. A brief, modified version of this taxonomy is included here to aid you in developing and writing affective objectives.[4]

I. *Receiving* (the ability to demonstrate an awareness of and a passive willingness to receive or attend to stimuli presented)

A. Awareness
(ability to demonstrate consciousness of a certain situation, phenomenon, object, or state of affairs)

B. Willingness to receive
(ability to demonstrate a tolerance for a given stimulus)

C. Controlled or selected attention
(ability to demonstrate controlled attention for a given stimulus despite competing and distracting stimuli)

II. *Responding* (ability to demonstrate a low level commitment to the stimuli presented)

A. Acquiescence in responding
(ability to demonstrate a compliance in reacting to a given stimulus)

B. Willingness to respond
(ability to demonstrate voluntary responses in reacting to given stimuli)

C. Satisfaction in response
(ability to demonstrate a feeling of pleasure, or enjoyment in responding to given stimuli)

III. *Valuing* (ability to demonstrate consistent behavior that reflects internalized values)

A. Acceptance of value
(ability to demonstrate acceptance of a value or belief)

B. Preference for a value
(ability to demonstrate behavior that reflects the pursuing, seeking, or wanting of a value)

C. Commitment
(ability to demonstrate a deep involvement with, and commitment to, a belief or value)

[4]From *TAXONOMY OF EDUCATIONAL OBJECTIVES: The Classification of Educational Goals: HANDBOOK II: Affective Domain,* by David R. Krathwohl et al. Copyright© 1964 by Longman Inc. Reprinted by permission of Longman.

IV. *Organization* (ability to demonstrate behavior that is reflective of an internalized, organized value system)

   A. Conceptualization of a value
      (ability to demonstrate behavior that reflects a relationship between old values held and new values being developed)

   B. Organization of a value system
      (ability to demonstrate behaviors that reflect an integrated and ordered relationship among values held)

V. *Characterization by a value or value complex* (ability to consistently demonstrate behavior reflective of an internalized value system to such an extent as to be described and characterized by it)

   A. Generalized set
      (ability to demonstrate behaviors that reflect a persistent and consistent response to related situations or objects)

   B. Characterization
      (ability to demonstrate a broad range of behaviors that are reflective of one's philosophy of life)

These five categories are hierarchical with Category I—Receiving—identifying the lowest level of affective objectives and Category V—Characterization by a Value or Value Complex—identifying the highest level. They also form a continuum on which each succeeding category represents a higher level of internalization. Although there is still much controversy and debate among educators as to the extent to which the school should be concerned with affective objectives, few deny that affective behaviors shape and determine to a great extent the learning outcome. It would seem that teachers cannot afford to avoid this area of learning, but, rather, must examine it in order to determine what is desirable and necessary.

Summary

This section has provided you with some background information related to developing and writing objectives. Objectives are used to give purpose and direction to instruction. To be most useful, they should be stated in terms of observable student behavior. Certain words, such as *state, explain, demonstrate, compare,* and *hypothesize,* have proven useful in writing behavioral objectives. However, just stating objectives so that they describe observable learner behavior does not assure that an objective is significant and valuable.

There are four main behavior categories into which objectives can be grouped: cognitive content, cognitive process, psychomotor, and affective. Effective science instruction involves all four of these categories. Modified taxonomic outlines of the three domains—cognitive, psychomotor, and affective—were included in this section to provide guidance in developing and writing significant science objectives. On the following pages, you will find activities designed to help you assimilate and accommodate the information presented thus far. You may want to refer to this section while doing the activities.

Activities

A. Work with a small group to identify an appropriate science topic for children in one of the following categories:

1. Kindergarten-Grade 1
2. Grades 2-4
3. Grades 5-6
4. Grades 7-8

It might be helpful to review the materials examined in Unit 2 for ideas. Make sure your group has enough background information about the selected topic to deal effectively and knowledgeably with the content presented. See your instructor if you need more guidance in selecting an appropriate science topic.

B. Using the ideas and information gathered by examining the various resources available, work with your small group to formulate appropriate behavioral objectives in each of the following behavior categories:

1. Cognitive content
2. Cognitive process
3. Psychomotor
4. Affective

You might find it helpful to think first in terms of instructional objectives—What are you trying to accomplish in each of the behavior categories identified above? Once you have identified your instructional purposes, it is usually not too difficult to use behavioral terms.

C. Use the following format to organize your instructional objectives:

1. *Cognitive content:* I want the students to know that

2. *Cognitive process:* I want the students to develop or improve their ability to

3. *Psychomotor:* I want the students to develop or improve their ability to

4. *Affective:* I want the students to

D. Using the above instructional objectives, convert them to observable student behavior. Ask yourself, "What do the students have to *do* to show me they have met my objectives?" Use the following format to organize your behavioral objectives:

1. *Cognitive content:* The students will

2. *Cognitive process:* The students will

3. *Psychomotor:* The students will

4. *Affective:* The students will demonstrate

    by

SELF-CHECK √

1. Look over each of your instructional objectives to see if you selected valuable, worthwhile goals that provide direction and purpose for instruction. Can you give some rationale for each of your choices? Discuss the rationale with your small group so that you can share this information with your instructor during the small group seminar.
2. Examine your behavioral objectives:
    a. Do they project specific outcomes of learning in terms of observable student behavior?

b. Is it clear what the students must *do* to let you know they have met each objective?

c. Do the behavioral objectives relate directly to the instructional objectives?

Jot down any comments or questions you want to share with your instructor during the small group discussion.

COMMENTS

3. When you have completed the section "Developing and Writing Objectives," invite your instructor for a seminar to discuss your progress.

NOTES

BIBLIOGRAPHY

Bloom, Benjamin S. et al. *Taxonomy of Educational Objectives, Handbook I: Cognitive Domain*. New York: Longman Inc., 1956.

Carin, Arthur, and Sund, Robert. *Teaching Science through Discovery*. 3d ed. Columbus, Ohio: Charles E. Merrill Publishing Co., 1975.

George, Kenneth. *Elementary School Science Why and How*. Lexington, Mass.: D.C. Heath Co., 1974.

Harrow, Anita J. *A Taxonomy of the Psychomotor Domain*. New York: Longman Inc., 1972.

Krathwohl, David R.: Bloom, Benjamin S.; and Masia, Bertram B. *Taxonomy of Educational Objectives, Handbook II: Affective Domain*. New York: Longman Inc., 1964.

Mager, Robert F. *Developing Attitudes Toward Learning*. Belmont, Calif.: Fearon Publishers, 1968.

Mager, Robert F. *Goal Analysis*. Belmont, Calif.: Fearon Publishers, 1972.

SUGGESTED READING

Bloom, Benjamin S. et al. *Taxonomy of Educational Objectives, Handbook I: Cognitive Domain*. New York: Longman Inc., 1956.

Carin, Arthur, and Sund, Robert. *Teaching Science through Discovery*. 3d ed. Columbus, Ohio: Charles E. Merrill Publishing Co., 1975. Chapter 4.

Combs, Arthur W. et al. *The Professional Education of Teachers: A Humanistic Approach to Teacher Preparation*. 2d ed. Boston: Allyn and Bacon, 1974.

George, Kenneth. *Elementary School Science Why and How*. Lexington, Mass.: D.C. Heath Co. 1974. Chapter 3.

Harrow, Anita J. *A Taxonomy of the Psychomotor Domain*. New York: Longman Inc., 1972.

Krathwohl, David R.; Bloom, Benjamin S.; and Masia, Bertram B. *Taxonomy of Educational Objectives, Handbook II: Affective Domain*. New York: Longman Inc., 1964.

Maslow, A.H. *Toward a Psychology of Being*. New York: Van Nostrand Reinhold, 1954.

Rogers, Carl R. *Freedom to Learn*. Columbus, Ohio: Charles E. Merrill Publishing Co., 1969.

Selecting and Developing Appropriate Teaching Techniques

Background Information

This section continues the planning process by offering some guidelines to follow in selecting appropriate teaching techniques. Teaching techniques or methods are used to aid the learner in attaining the specified objectives.

The ideas of both Jean Piaget and Jerome Bruner provide insight for the elementary teacher. Piaget's research suggests that certain factors are involved in promoting learning:

1. *Physical experience*—manipulation of real materials
2. *Social experience*—interaction with others, confronting views and ideas of others
3. *Logical-mathematical experience*—activities involving bringing together, taking apart, grouping, counting
4. *Maturation*—the passage of time.
5. *Equilibration*—a mental state of balance[5]

The fifth factor listed is considered by Piaget to be the most essential. He asserts that the human organism strives for equilibration. When something occurs that upsets the existing mental structure, there is a basic tendency to restore it. Our mental

This teacher is using physical experience as one teaching technique.

---

[5]Hans G. Furth, *Piaget for Teachers* (Englewood Cliffs, N.J.: Prentice-Hall, 1970).

structure is built over the years through the different experiences we have in which data are collected and incorporated into the existing framework or used to modify the existing framework. When new data are encountered that fit into the existing framework, according to Piaget, *assimilation* takes place. On the other hand, if data are encountered that do not fit into the existing mental structure, modification must be made in the existing structure to *accommodate* the new data. Thus, the mechanisms of assimilation and accommodation become essential in building a mental structure.

It is important to note that Piaget's five factors are considered necessary but not sufficient; meaning that just providing a child with these factors will not automatically result in learning, but without them learning cannot take place. Essentially, it is the *interaction,* or active involvement of the child with the environment, that causes learning to take place.

These five factors can be used to guide teachers in developing appropriate teaching strategies, methods, and techniques. Children should be exposed to a physical and social learning environment that provides them with opportunities to manipulate objects and ideas in active interaction with others. Freedom of move ment,individualized and personalized instruction, internal motivation, flexible curriculum, a variety of relevant concrete experiences, and direct involvement are important elements to consider in developing teaching strategies.

In addition, the element of time needs to be put into proper perspective. Piaget sees the passage of time as necessary for assimilation and accommodation to take place. Teachers often try to speed up the learning process, but frequently this results in false or verbal accommodation and not true internalization of the concept or process. Thus, a child may seem to have grasped the idea but only at a memorization level. Therefore, time must be allowed for the students to interact directly with objects, ideas, and other people.

Bruner thinks that students learn best by discovery.[6] He suggests that allowing the learner to discover information and organize what is encountered is a necessary condition for learning the techniques of problem solving. This philosophy is consistent with Piaget's emphasis on active learner involvement.

Bruner offers guidance in developing instructional strategy by suggesting there are three modes of presentation: action, imagery, and language.

The *action* mode provides actual contact with real objects. Children are presented with objects to manipulate in order to acquire the desired learning outcome. This mode of presentation is considered most appropriate for learners in Piaget's preoperational and concrete operational stages of mental development. It is also a useful mode for learners in any stage who are encountering a new concept for which they have little background.

The *imagery* mode uses pictures, models, diagrams, and other such representations of actual objects to foster a desired learning outcome. Learners who are making the transition from concrete to formal thinking are able to utilize the imagery mode of presentation in acquiring desired learning outcomes. This mode is not generally as effective with learners at earlier stages of mental development.

When a learner reaches what Piaget terms the formal stage of mental development, the *language* mode of instruction can be utilized to produce a desired learning outcome. Words, abstractions, and other symbols characterize this mode of instruc-

---

[6]Jerome Bruner, *Toward a Theory of Instruction* (Cambridge, Mass.: Harvard University Press, 1967.)

Bruner's three modes of presentation are action, imagery, and language.

tion. Experiences involving the use of the other two modes serve as necessary prerequisites for the language mode.

It can be inferred, then, that in planning for instruction the different modes of presentation need to be considered. The stage of development at which a child is operating needs to be known and plans made accordingly. Since the stage of development is not wholly dependent on age but is influenced also by past experiences, the task of selecting appropriate instructional methods or techniques is not simple. It would seem that several paths need to be provided by which learners can reach the desired outcome.

The work of Piaget and Bruner offer the classroom teacher valuable insight and guidance in planning for science instruction. Two types of planning are necessary: daily and long range. Long-range plans frequently are organized around a selected broad topic or concept in science. The teacher selects and organizes instructional materials and activities that will help the students develop the major ideas, concepts, and skills that relate to the selected topic. Daily planning is a necessary part of long-range planning, and long-range planning provides the structure for daily planning.

The focus of this Part is to help you make effective *daily* plans, but the ideas presented also offer guidance in making long-range plans. The following are four vital elements that should be a part of all daily instruction:

1. Attention-getting and motivating techniques
2. Data-gathering techniques
3. Data-processing techniques
4. Closure techniques

Attention-getting and Motivating Techniques

To begin a lesson, it is obvious that you must first get the students' attention. Also, you want to motivate them so they feel a need to continue with the instruction. There are several techniques that can be used in getting students' attention and motivating them to become active learners. Piaget's theory on equilibration provides one avenue. Discrepant events or inconsistencies are valuable instructional techniques that cause the learner to become disequilibrated. This means that information is presented that does not fit in with the learner's existing mental structure. As a result, the learner becomes uncomfortable or frustrated. Usually this produces a desire on the part of the learner to resolve the inconsistency.

A discrepant event can only be an effective technique if it relates to a learner's past experiences. That is, a learner must have some background to enable him to perceive the event as discrepant or inconsistent. Because of previous experiences, the learner has acquired a set of expectations that prove unreliable when the new event occurs. To use Piaget's terms again, the learner is not able to *assimilate* the new information and must modify her existing mental structure to *accommodate* it. By using this instructional technique to begin a lesson, a teacher is able to capitalize on the learner's natural curiosity and the need to resolve the inconsistency. The remainder of the instructional sequence would be used to provide the learner with necessary materials and guidance to resolve the inconsistency.

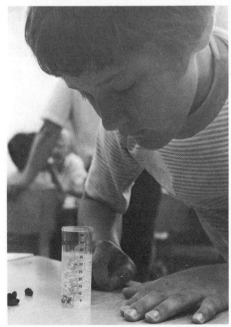

This student is motivated by his desire to resolve a discrepancy presented by the experiment.

There are many ways to present a discrepant event. Kenneth George has suggested the following:

1. Silence—use exaggerated movements and conduct the entire presentation of the discrepant event in silence.
2. In progress—have the event already in progress when student enters the teaching area.

3. Pictures and film loops—use visual aids that show inconsistent observations or present information that is inconsistent with past experiences of the learner.[7]

If you select a discrepant event to use for the attention-getting and motivating portion of the instructional sequence, it is important to keep a few things in mind. Make sure that the event is *directly* related to the rest of the instruction. The data-gathering and data-processing portions of the instruction should involve the child with appropriate materials and guidance in resolving the inconsistency created.

A second consideration that must be given attention involves the frequency of using the discrepant event technique. For best results, it is better to limit the usage of this technique. Children sometimes begin to expect an entertainer instead of a teacher if the technique is used too frequently. Or they begin to expect the unexpected, thus the technique loses its purpose.

Discrepant events do provide an excellent beginning for a lesson, but as you now realize, a teacher must use a variety of techniques to be most effective. One of the most used beginning techniques is *verbal discussion.* The teacher begins the lesson by establishing background information through a lecture or question-answer session in order to relate to the students' past experiences. This information is then related to the present instruction planned for that day's lesson. The students are prepared during this portion for what is to come during the data-collecting and data-processing portion of the lesson. This technique is not as exciting as a discrepant event, but it does serve to get the students' attention and induce some motivation. For example, to begin a lesson on magnets, the teacher might begin by holding up a horseshoe magnet and asking the children to identify it. Further discussion would continue and opportunities would be given to the students to express their personal experiences with and knowledge of magnets. The teacher would guide the discussion so that the students could establish connections with past experiences that would be helpful in relating to the new information.

This technique can be expanded to include a *problem presentation.* For example, the teacher could guide the discussion so that a problem situation is presented. In the previous discussion about magnets, the teacher could ask a question such as, "Do you think a magnet can attract any object? Let's find out." Of course, the kind of problem created would depend on the stage of mental development of the students and their past experiences with the topic area. The problem presented must also be one that can be solved by the students. Sometimes *conflicts of opinion* that arise during the verbal discussion can be used effectively. It's best if these conflicts have more than one correct solution. Hence, you should avoid having students engage in competition with one another if the end result is a winner and a loser. But you can create a situation during the verbal discussion which fosters competition that will result in alternative means and solutions.

Combinations or adaptations of any of the techniques presented here can be used in the beginning portion of a lesson to get the students' attention and motivate them to continue with the rest of the instructional sequence.

Data-gathering Techniques

Once you have gotten the students' attention and motivated them, you are ready to move into the next portion of the lesson—involving them in gathering data. You are

---

[7]Reprinted by permission of the publisher, from George Dietz et al., *Elementary School Science: Why and How* (Lexington, Mass.: D.C. Heath and Company, 1974).

now ready to provide the eager students with some experiences that focus on collecting information. The information collected should relate directly to specific situations created by using the attention-getting and motivating techniques. The three modes of presentation—action, imagery, and language—can be used separately or, probably more effectively, in combination in providing a format for data-collecting activities.

The action mode calls for firsthand experiences in which the students manipulate the objects in order to gather data. Engaging the students in a variety of laboratory activities is an excellent way to accomplish this. Laboratory activities can involve the students in simple teacher-directed tasks or can provide an outlet for creative problem solving. Sophisticated equipment is not needed for purposeful learning to take place. Simple everyday items can be used effectively to provide learners with firsthand experiences in data collecting. Furthermore, laboratory activities can take place outside the classroom as well as inside. Field trips on the school grounds can provide an excellent opportunity for students to be involved in firsthand experiences. Part 2-1, "Laboratory Approach to Elementary Science," involved you in a variety of action mode activities. You can refer to it for ideas during the activity section.

Data-gathering techniques using the imagery mode of presentation involve the students with representations of real objects. Models, pictures, films, charts, or other visual aids are typical examples. As stated earlier, however, the student needs previous concrete experiences to profit from the imagery mode of presentation. Generally, children who are in the transitional stage between concrete operations and formal operations can benefit more from this mode of presentation than can those at lower levels of cognitive development. The imagery mode of presentation is very useful in providing students with the opportunity to examine data not ordinarily available to them. Plastic models of body organs and the solar system are examples.

*Field trips are excellent
data-gathering activities.*

The language mode of presentation is the most abstract of the three modes and is most appropriately used with students who have reached the formal stage of cognitive development. Lectures and reading assignments are typical examples of data-gathering techniques in this mode. Information obtained in this manner is considered to come from a secondary source. Secondary sources are legitimate sources of scientific information, but they do differ from firsthand manipulations of equipment and materials. Overuse of language mode activities results in a product oriented science approach. Remember that prior experience in the action and imagery modes are necessary prerequisites for effective functioning in the language mode.

Working in a group to organize data into a chart is a good data-processing activity.

Data-processing Techniques

Data-processing techniques are used to involve the learner in organizing and analyzing the data gathered. In data gathering, the learner quite often works alone; in data processing, interaction among the learners is encouraged. Group discussion, which involves a sharing and comparing of information gathered, is a good data-processing technique. Some questions that lead students to analyze information are these: "What patterns do you see developing?" "Why do you think this happened?" "Can you support that statement?" "Why do you say that?" Other questions can be used to help promote peer interaction: "Brenda, do you agree with Glen's interpretation?" "Bob, can you add anything to Lynn's explanation?"

Organizing the data gathered into chart or graph form is a good data-processing activity. Making predictions, formulating hypotheses, making inferences, and classifying are other examples of data-processing skills. In all of these examples, the learner is involved in doing something with the data gathered. It may involve physical manipulations of the data, as in constructing charts and graphs, or mental manipulations, as in predicting or formulating hypotheses. Providing opportunities for learners to engage in data processing is an essential part of all science instruction. Effective data-processing techniques promote assimilation and accommodation.

To begin the lesson, you aroused the students' interest with attention-getting and motivating techniques. Next, you provided experiences that involved them in collecting data. Once the data were collected, you guided the students in organizing and analyzing it, which led to assimilation and accommodation. You are now ready for the final phase—closure.

Closure Techniques

Helping the learner reach closure is an important part of any lesson. The teacher must help the learners view the total learning experience. She must help bring into focus the various parts of the instruction and establish relationships.

A summarizing discussion is quite valuable in reviewing the tasks completed and pulling everything together in proper perspective. A question-and-answer format can be utilized to involve the learners directly. On the other hand, in some cases a teacher monologue is more appropriate. Any generalizations made and conclusions reached should be related to the beginning sequence of the lesson. If the attention-getting and motivating technique used created an inconsistency or a problem, make sure that the situation is clear for the learner.

Summary

Teaching techniques are the methods of instruction used by teachers to aid the students in attaining specified objectives. Research results indicate that children should be exposed to a physical and social learning environment that provides them with opportunities to manipulate objects and ideas in active interaction with others. Teaching strategies which provide this kind of learning environment demand that students be given time to assimilate and accommodate the new experience.

According to Bruner, instructional strategy can be categorized as action mode, imagery mode, and language mode. These three modes of presentation can be used separately; however, they would probably be more effective in combination in providing a format for science activities.

All daily science instruction should involve four elements: (1) attention-getting and motivating techniques, (2) data-gathering techniques, (3) data-processing techniques, and (4) closure techniques. To begin a lesson, you must first get the students' attention, then motivate them so they feel a need to continue with the instruction. There are various techniques that can be used to accomplish this task. Discrepant events provide an excellent beginning for a lesson, provided they aren't overused. A verbal discussion that prepares the students for instruction by giving a preview of what is to come is also a useful technique. An opening discussion can also be used to present a problem to be solved or create a situation that fosters competition that will result in students' using alternative means to find solutions.

The next portion of the lesson involves the learner in data gathering. Laboratory activities provide an excellent means of involving students with firsthand experience in data gathering. These activities can take place inside the classroom or outside as part of a field trip. The use of models, pictures, films, charts, or other visual aids provides the students with the opportunity to examine data they cannot experience directly. Information can also be obtained from secondary sources, such as reading and lectures. But, overuse of this technique can result in a content- or product-oriented science approach.

Once the data are collected, the teacher uses data-processing techniques to involve the students in organizing and analyzing the information. Guided discussion is useful in promoting and encouraging interaction among the learners. Engaging the students in both physical manipulation of the data—constructing charts and graphs—and in mental manipulation of the data—predicting, inferring, etc.—is an essential part of all science instruction.

Developing techniques that help students reach closure is the final phase of instruction. A teacher monologue which summarizes the lesson or a question-and-answer format in which students participate can be used to help learners put the lesson in proper perspective.

The activities that follow have been designed to help you utilize the ideas presented here. You will work with a small group in selecting and developing appropriate teaching techniques that could be used to aid the learner in attaining the objectives formulated in the preceding section. You will probably want to refer to portions of the background information for guidance in developing your instructional strategy. Be sure to talk with your instructor if you need more clarification and guidance in understanding the material or in completing the activities.

A. Working with your small group and using the objectives you developed in *"Developing and Writing Objectives,"* select the appropriate teaching techniques and develop an instructional sequence that could be used to help students attain the specified objectives. You will probably want to read over your responses and comments from the preceding activities to refresh your memory. Record the grade level selected and the behavioral objectives written earlier. You may find it necessary to modify your original objectives in order to develop more effective instructional strategy.

Grade Level:

Behavioral Objectives:

1. *Cognitive content:* The student will

2. *Cognitive process:* The student will

3. *Psychomotor:* The student will

4. *Affective:* The student will demonstrate

   by

B. Using the ideas and information presented in the background information, develop an instructional sequence that could be used to aid the learner in obtaining your specified objectives. Use the format below to organize and describe your instructional strategy.

Instructional Strategy:

1. Attention-getting and motivating techniques:

2. Data-gathering techniques:

3. Data-processing techniques:

4. Closure techniques:

SELF-CHECK √

A. Examine your instructional strategy to see if it provides the learners with enough appropriate practice to enable them to attain the specified objectives. Evaluate your instruction using the following criteria:

1. The instruction will enable the learners to perform the behavioral tasks specified in the behavioral objectives.
2. The instruction provides appropriate practice in each of the four behavior categories—cognitive content, cognitive process, psychomotor, and affective.
3. The instruction incorporates the ideas of Piaget and Bruner by providing for (a) manipulation of real objects, (b) interaction with others, (c) guided discovery, and (d) alternate learning paths.
4. The attention-getting and motivating technique relates to the learners' past experience and provides a reason for participating in the learning experience.
5. The data-gathering technique involves the learners with experiences that focus on collecting information to be used in dealing with the specified situations created in the attention-getting and motivating portion of the instruction.
6. The data-processing technique involves the learners in the following:
   a. Organizing and analyzing the information gathered
   b. Interacting with classmates
   c. Assimilating and accommodating
7. The closure technique helps the learners to pull it all together. All parts of the instruction are related and seen as a whole. A state of equilibrium is achieved by the learners.

COMMENTS

B. After you have discussed this activity with your small group, invite your instructor to join you for a brief seminar.

NOTES

BIBLIOGRAPHY

Alney, M., Chittenden, E. A., and Miller, P. *Young Children's Thinking: Studies of Some Aspects of Piaget's Theory.* New York: Teachers College Press, 1966.

Bruner, Jerome S. *Toward a Theory of Instruction.* Cambridge, Mass: Harvard University Press, 1967.

Bruner, Jerome S., Goodnow, J. J., and Austin, G. A. *A Study of Thinking.* New York: John Wiley & Sons, 1956.

Furth, Hans G. *Piaget for Teachers.* Englewood Cliffs, N.J.: Prentice-Hall, 1970.

George, Kenneth. *Elementary School Science: Why and How.* Lexington, Mass.: D. C. Heath Co., 1974.

Piltz, Albert, and Sund, Robert. *Creative Teaching of Science in the Elementary School.* 2d ed. Boston, Mass: Allyn and Bacon, 1974.

SUGGESTED READING

Carin, Arthur, and Sund, Robert. *Teaching Science through Discovery.* 3d ed., Columbus, Ohio: Charles E. Merrill Publishing Co., 1975. Chapter 7.

George, Kenneth. *Elementary School Science: Why and How.* Lexington, Mass.: D. C. Heath Co., 1974. Chapters 5-9.

Utilizing Evaluation Techniques

Background Information

Evaluation refers to the difficult task of assessing the extent to which the students were able to attain specified objectives. If the objectives are stated in terms of observable student behavior and are used to provide the framework for developing appropriate instruction, then the task of assessment is much easier.

Two types of evaluation are considered a necessary part of all science instruction—*formative* and *summative*. Formative evaluation takes place *during* the instructional sequence and is used to diagnose a learner's needs. This type of evaluation provides immediate feedback that is used to guide the learner in completing the task at hand. Summative evaluation takes place *after* the instruction has terminated and is used primarily as a basis for assigning grades.[8]

Both types of evaluation can take various forms, but generally there are three broad categories that can be used to encompass them.

1. *A paper-and-pencil test* is probably the most widely used form of science assessment. It is used primarily for summative evaluation but can be used equally as well for formative evaluation.
2. *A project or written report* is another frequently used form of evaluation, particularly in the upper elementary grades. This form works well when your objectives are application oriented.
3. *The performance task* is relatively new as an evaluation technique but offers much promise in assessing science learning. Essentially, the learner is provided with the necessary materials and is observed performing the specified task. This form of evaluation lends itself to the process-inquiry approach of sciencing.

Paper-and-Pencil Test

There are several different types of paper-and-pencil tests. No one type has proven more successful than any other, but rather it is the matching of the test format with the learning objectives to be assessed that has proven critical. Before selecting any instrument to assess a student's achievement, you must be able to identify what it was that the student was supposed to accomplish. This brings us back to objectives. Objectives become the binding thread that holds together and integrates an entire lesson. The objectives must be examined and used as a guide in the selection of an assessment instrument.

There are six popular types of paper-and-pencil tests: multiple choice, fill in the blank, true-false, matching, short answer, and essay. All of these types involve

---

[8]Benjamin Bloom et al., *Handbook on Formative and Summative Evaluation of Student Learning* (New York: McGraw-Hill, 1971).

recalling information and identifying a correct response. In some types, such as multiple choice, true-false, and matching, very little writing is required. The correct choice is indicated by circling, marking, drawing a line, or perhaps writing a letter or number. These tests are usually considered to be objective, meaning that there is one correct answer, and can be scored using a key which contains all the correct answers.

The following are some examples of objective tests:

A. *Multiple Choice.* Circle the correct response.

   1. All matter is composed of tiny particles called:

      a. elements

      b. molecules

      c. compounds

      d. gases

   2. The pressure of liquids or gases will be low if they are moving fast and will be high if they are moving slowly.
     This principle is called:

      a. Newton's principle

      b. Galileo's principle

      c. Bernoulli's principle

      d. Jenner's principle

B. *True-False.* Circle the correct response.

   Ⓣ   F   1. Limestone and marble are chemically the same.

      T   Ⓕ   2. The point where the earth's crust cracks and moves is called a geyser.

C. *Matching.* Draw a line to connect the word on the left with the appropriate example on the right.

    a. mammal   1. frog

    b. amphibian   2. dog

    c. reptile   3. ant

    d. bird   4. lizard

    e. insect   5. chicken

The other paper-and-pencil tests mentioned—short answer, fill in the blank, and essay—require more grading time, frequently have a variety of acceptable responses, and tend to be more subjective in nature. Students are generally required to do more recalling as well as organizing the stored information if these types of tests are used. The simplest of these three is the fill in the blank. Sometimes there is only one word missing, but the student must recall it without benefit of being presented with four from which to choose. More often, several words are omitted, and the student must supply the missing thought. Communicating the general idea, not exact wording, is the issue.

The short answer and essay form of evaluation expand on the fill in the blank type. Generally, a question is asked or a term given, and the learner must define, describe, explain, and give examples in answering the question. This type of evaluation is most appropriate for children who are toward the end of the concrete stage of mental development or already operating at the formal level. Children at lower levels of cognitive development have not sufficiently developed the skills necessary to organize and express their thoughts through writing.

Following are some examples of subjective tests:

A. *Fill-in-the-blank*

1. Most plants have four parts: *(roots)*, *(stems)*, *(leaves)*, and *(flowers)*.
2. One of the main purposes of the stem of a plant is (*to carry water from the roots to the leaves*).

B. *Short Answer*

1. How does a rock differ from a plant?
2. What causes food to spoil?

C. *Essay*

1. Describe and compare two ways in which a plant can reproduce.
2. How does the environment affect living things?

Paper-and-pencil tests do provide an effective means of assessing learner achievement. A variety of forms provides the teacher with a choice as to which form or combination of forms could best be used in evaluating the learning process. Paper-and-pencil tests should not limit the learner to low-level memory responses. The recall of facts and information is certainly a necessary part of all learning, but it must not be all that is required. Higher levels of cognitive functioning should be elicited also. You might want to turn back to page 195 and review Bloom's levels of cognition. These levels offer guidance in formulating test items that elicit responses at higher levels of cognition.

Although most paper-and-pencil tests are used as summative evaluation to provide a basis for grading, they can be used effectively as formative evaluation. The format of the test remains the same, but in summative evaluation, the results are not used to plan for further instruction as is the case with formative evaluation. Thus, the main difference between the two forms of evaluation is in the use of the test results. In formative evaluation, the results become a part of the instructional process and students are able to use them. In summative evaluation, the test results are not used because essentially the learning process is over, a grade is given, and the class proceeds to another area of learning.

Projects and Written Reports

A project or written report is appropriate to use in assessing a students' achievement if the objectives are more application oriented. This type of evaluation usually involves students in long-term preparation. The final product is presented and the student's achievement is measured by comparing the project results and/or the written report with the established criteria. If the students and teacher are in constant communication throughout the project and feedback is given at several stages, this type of evaluation can be formative as well as summative.

Many science teachers find this type of evaluation more appropriate than paper-and-pencil tests. It allows for assessment of the skills of process-inquiry oriented science and isn't limited to mental manipulation of the concepts.

Performance Tasks

The performance task is closely related to the project and written report type of evaluation. Generally, it doesn't require as much time as does a project or written report. The students are provided with the necessary materials and equipment and demonstrate achievement of the specified objectives by performing an appropriate task. The teacher observes and evaluates using criteria based upon the specified objectives.

This type of evaluation is particularly appropriate for assessing attainment of cognitive process, psychomotor, and affective objectives. Use of a checklist or other method for recording observations is helpful in using this type of evaluation.

The performance task provides an excellent means of formative evaluation. The students are involved in both the means and end of instruction. As the instruction is taking place, the students are involved in appropriate practice activities and receive feedback and guidance from the teacher. This appropriate practice also becomes the end product of instruction as the students perform the task and are observed for the purpose of making a final assessment of the performance.

Summary

Evaluation is an essential part of all science instruction. It has two functions:

1. *Formative*—providing feedback during instruction and offering guidance for continued learning in the same area
2. *Summative*—providing a basis for assigning a grade after instruction is completed

Assessing the extent to which the students have attained the specified objectives of instruction is a difficult task. A variety of techniques are needed, which a teacher can select, adapt, and modify to fit the requirements of a specific situation.

Paper-and-pencil tests, projects and written reports, and performance tasks form three broad categories of evaluative techniques into which the various forms of each can be grouped. The selection of an evaluation instrument involves a matching of the instrument with that which is being assessed. In order to make an appropriate match, the teacher must use the stated objectives and the instructional procedure as guides.

The following activities are designed to help you utilize the ideas presented here. You may want to talk with your instructor at this point if you need help in understanding the information presented so far.

A. This activity will involve you in the total planning process. In the section, "Developing and Writing Objectives," you developed and wrote instructional and behavioral objectives. In "Selecting and Developing Appropriate Teaching Tactics," you used those objectives to develop an instructional sequence which included attention-getting and motivating techniques, data-gathering techniques, data-processing techniques, and closure techniques. In this final phase of planning, you are to develop appropriate evaluation techniques that can be used in assessing the extent to which the learner has attained the specified objectives you developed previously.

B. Work with your small group and examine the first two phases of the planning sequence. You may find that you wish to alter or modify the original objectives and/or instructional strategy now that you can view the entire planning process from an integrated point of view. In fact, you may decide to discard your original efforts altogether and begin anew. Before making the decision, take time to review all three sections of this Part, both the reading sections and the corresponding activities. When you have made a decision, record your group plan on separate sheets to hand in to your instructor during the instructor seminar. Use the following outline to describe your entire lesson plan.

Title of Lesson:

Grade Level:

Behavioral Objectives:

1. *Cognitive content:* The learner will

2. *Cognitive process:* The learner will

3. *Psychomotor:* The learner will

4. *Affective:* The learner will demonstrate

by

Instructional Strategy:

1. Attention-getting and motivating techniques:

2. Data gathering techniques:

3. Data-processing techniques:

4. Closure techniques:

Evaluation:

1. Technique selected:

2. Example of assessment:

   a. If the instrument is a paper-and-pencil test, develop the actual test and provide the correct answers. Indicate what percent of the answers should be answered correctly before you can say that the students have met the objectives.
   b. If the instrument is a project, written report, or performance task, provide a complete description of the procedure to be followed. Include both written and verbal directions. A checklist that lists criteria to be used in judging the student's performance should also be included.

√ SELF-CHECK

1. When your group lesson plan is completed, use the following questions to guide the group discussion and evaluation of this module:

   a. What were the main learning theories and ideas of Piaget and Bruner that were presented in the module?
   b. What implications do these theories have for constructing teaching strategies?
   c. How are the ideas of Piaget and Bruner reflected in your lesson plan?

2. Turn back to pages 201-202 and use the criteria presented there to evaluate the objectives developed for your final lesson plan. Make any adjustments you feel are necessary.

3. Use the criteria on page 215 to evaluate your instructional strategy. Make the necessary modifications that result from this evaluation or that result from changes in the objectives.

4. Use the following questions to judge the effectiveness of the evaluation portion of your lesson plan:

   a. Are *all* the objectives—cognitive content, cognitive process, psycho-motor, and affective—evaluated?

   b. Is both formative and summative evaluation included?

   c. Can you offer a convincing rationale for your choice of evaluation technique?

COMMENTS

BIBLIOGRAPHY

Bloom, Benjamin, Hastings, J. Thomas, and Madaus, George F. *Handbook on Formative and Summative Evaluation of Student Learning.* New York: McGraw-Hill, 1971.

Carin, Arthur, and Sund, Robert. *Teaching Science through Discovery.* 3d ed. Columbus, Ohio: Charles E. Merrill Publishing Co., 1975.

George, Kenneth. *Elementary School Science: Why and How.* Lexington, Mass.: D. C. Heath Co., 1974.

SUGGESTED READING

Carin, Arthur, and Sund, Robert. *Teaching Science through Discovery.* 3d ed. Columbus, Ohio: Charles E. Merrill Publishing Co., 1975.

George, Kenneth. *Elementary School Science: Why and How.* Lexington, Mass.: D.C. Heath Co., 1974.

Summary: Planning for Teaching

In this Part you were involved in a step-by-step process of developing a teaching strategy which included instructional objectives, behavioral indicators, techniques or methods to be utilized, and evaluation to be employed. Review the goals and objectives for this Part listed on page 191. If you feel that you have successfully met these goals and objectives, arrange for a final seminar with your instructor. If not, ask your instructor for individual help.

FINAL SEMINAR

Discuss your planned teaching strategy with your instructor. Share the results of your group discussion with your instructor also. Together you may want to use the self-check to guide your final seminar.

NOTES

COMPETENCY EVALUATION

Your instructor may choose to use a competency evaluation measure of some type to evaluate your competency in the area of planning for teaching. Check with your instructor for specific directions.

PART 4-2

Questioning Techniques

FLOWCHART: Questioning Techniques

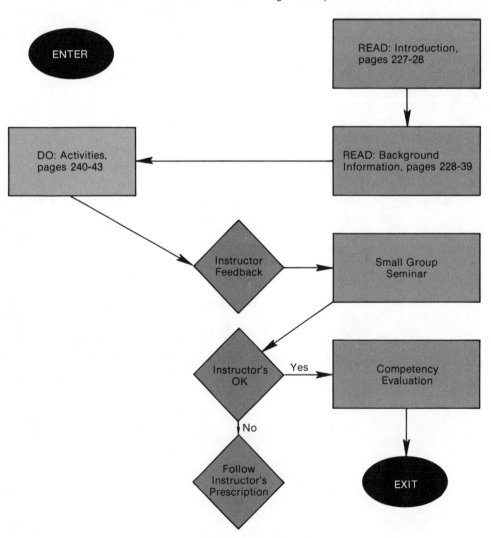

Introduction

The ability to formulate and utilize questions fostering creative, thoughtful, high-level responses is an integral part of the instructional strategy. Unfortunately, most classroom teachers ask questions that require only a low-level, memory response. Questions that encourage and solicit responses indicating more than rote memory need to be included in the instructional process if higher-level thinking is desired.

Effective questions require thoughtful responses.

There are various questioning strategies that offer guidance to elementary teachers in planning and integrating effective questions into the instructional strategy. In the background information section of this Part, the themes and ideas of two of the more well-known questioning strategies are presented. Information about the purpose and kinds of questions is also included. Factors to consider in designing your own questioning strategy for use with sciencing activities are also discussed.

Planning for effective questioning may involve a change in teacher-student relationships. This occurs because questions that require more than rote memory responses often stimulate students to respond with different answers from different points of view. The teacher must adjust to the new role of discussion facilitator. Furthermore, different kinds of questions demand different kinds of responses. A single "Yes Lynn, that's right," or "No Bob, that's wrong" is not an appropriate response for higher-level questions. The background information will provide some guidance in the area of responding to students' replies to higher-level questions.

Activities designed to involve you in formulating questions that foster creative, thoughtful, high-level responses are also included. In completing this Part, you may work alone in formulating the questions, but be sure you share the results of your efforts with your classmates and your instructor for feedback.

GOALS

After completing this Part, you will demonstrate competency in the following:
A. The ability to formulate effective questions that foster more than rote memory responses
B. The ability to utilize appropriate questioning strategies in formulating questions to be used as part of a planned science activity for elementary students

BEHAVIORAL OBJECTIVES

In completing this Part, you will do the following:
A. Identify and describe the questioning strategies presented in the background information section
B. Write science questions designed to elicit more than rote memory responses from elementary students
C. Utilize the ideas presented in this Part by selecting a science topic and constructing an appropriate questioning strategy as part of a planned science activity for elementary students

Background Information

Effective questioning is an art. Unfortunately, few teachers possess well-developed questioning techniques or strategies. Too often, they limit themselves to questions arising spontaneously during instruction. Carefully planned questions can offer direction and guidance for instruction. However, such effective questioning involves the use of specific skills, techniques, and strategies. Often a change in attitude toward the learning process itself is necessary.

In order to help you acquire the skills involved in effective questioning, it is necessary to examine the role of questioning in the learning process. Questions are most often used by teachers to see if the learner can remember information that was presented. This use of questioning is limiting, in that students are merely required to recall previously memorized information. Memory-level knowledge is considered to be the lowest level of cognition.

Questions can serve a more useful purpose than testing students' abilities to recall information. They can be used to stimulate student involvement and interaction. In Part 4-1, "Planning for Teaching" you used a sequenced format composed of four elements—attention-getting and motivating techniques, data-gathering techniques, data-processing techniques, and closure techniques—in making your plans for instruction. Effective questioning can be used in each of these phases to involve the students both mentally and physically in the learning process.

Questions for Stimulating Involvement

Questions designed to get students' attention and motivate them might focus on discrepancies, create a problem situation, result in conflicts of opinion, or aid the students in relating past experiences to new information. The following are examples of such questions:

1. How can it be that the water doesn't come out of the glass when it is turned upside down?
2. Does the length of the string affect the pendulum's period?
3. How can we find out which magnet is stronger?
4. Have you ever used a thermometer before? Where? What did you find out by using a thermometer? Is this thermometer like the one you have seen or used before? What could we find out by using a thermometer?

Questions for Data Gathering

Questions can also be structured to aid students in data collecting. In firsthand experiences, questions such as, "What does it feel like?" "Can you describe the sound it makes?" and "Does it have an odor?" guide the students in using their senses to make observations. Others focus the students' attention on quantitative measurements:"How heavy is it?" "How long is it?" Still others suggest actions the students might take in gathering information: "Do you think it will float?" "How high will it bounce?".

When data collecting is done through the use of pictures, models, charts, and graphs, questions such as the following can be used: "How would you describe the animal in this picture?" "According to the graph, what can you tell me about the rainfall in Michigan last year?" "After observing an operating model of our solar system, how would you describe the relationship in space of the planets to each other and to the sun?"

Questions that are designed to guide learners in gathering information from secondary sources, such as lectures and reading assignments, are often quite difficult to structure. Too often, questions merely involve the students in recalling information. Instead, they should serve to guide the learner in gathering the information. Consider the following examples: "In reviewing the reading assignment, how would you describe the process of oxidation?" or "After listening to the lecture on heat energy, how would you describe the ways in which heat energy travels?" Questions such as these can guide learners in gathering specific information that is presented in the language mode. Their purpose is *not* to have the learners recall from memory specific information but to have them use the source material in gathering the appropriate information. This same emphasis is reflected in the questions presented earlier which guide the students in using action and imagery sources in gathering data. By using questions with this emphasis, learners are encouraged to assume a more active role in the data-collecting process.

Questions for Data Processing

Following the data-gathering phase of instruction, the student is guided in organizing and analyzing the data. This phase of instruction, you will recall, is referred to as data processing. Questions used during this portion of instruction should require the learner to engage in one or more of the following process-inquiry skills: classifying, measuring, using space relationships, communicating, predicting, inferring, defining operationally, formulating hypotheses, interpreting data, controlling variables, and experimenting. All of these skills involve the learner in organizing and analyzing the data. Review the following sample questions and notice how they are designed to engage the learner in organizing and analyzing data.

1. How could you separate these objects into two different groups?
2. How can you present the data you gathered so that others can understand them?
3. Based on the observations you made about the object concealed inside the opaque container, can you identify it?
4. After examining the three closed, identical circuits, what reasons can you give for the bulb on only one of the circuits *not* lighting?
5. How would you determine the effect of the length of the pendulum on its period of oscillation?

Questions for Closure

Questions can also be used effectively in the final phase of the lesson—to help the learner reach closure. Questions that require the learners to relate the various parts of the lesson, cause them to focus on the main points, or serve to help them "put it all together" are most appropriate here. The following are some examples:

1. How do the kind, size, and length of materials used affect the flow of electricity?
2. Trey, our lesson today focused on making complete circuits. Can you look at this sample circuit and explain why it isn't complete?
3. What did we find out about the effect of the length of the pendulum on its period of oscillation?

You should now be familiar with the ways in which questioning can be used in each of the four phases of the lesson in order to involve the student both mentally and physically. The important thing to remember is the *purpose* of the question. Questions should be carefully chosen and worded so they contribute to the learning process in the way intended. Furthermore, careful planning can give a teacher the security needed to also include spontaneous questions.

Convergent and Divergent Questions

Another role that questioning plays in the learning process is that of stimulating convergent or divergent thinking. *Convergent questions* involve the learner in centering. Such questions have a limited number of acceptable or appropriate answers. In contrast, *divergent questions* have many acceptable responses. The students are encouraged to offer alternative solutions. Divergent questions also encourage creativity and require more than rote memory responses.

You need to plan for both convergent and divergent questions. The trick is to know when to use each type of question. Generally, divergent questions work best with the first three phases of a science lesson. In these phases—attention-getting and motivating, data-gathering, and data-processing—divergent questions allow for maximum input from students and foster interactions among them.

Convergent questions are most appropriate to use during the final phase of the lesson. In formulating a response to a convergent question, the students focus on specific bits of information. Their thinking is drawn in toward a common point. However, there may also be times when convergent questions are appropriate in the earlier part of the lesson. Careful planning will help the teacher to use the appropriate kind of question. Read over the following questions and see if you can distinguish the ones designed to encourage convergent thinking from those designed to elicit divergent thought:

1. Do your hands feel warm when you rub them together very rapidly?
2. What causes the molecules of a liquid to move?
3. How can we find out if heat affects the time it takes for a substance to dissolve in water?
4. How do you know air is around you?
5. Do you think the sugar cube will dissolve faster in the cold water than in the hot water?
6. What do you think will happen when the sugar is heated?

Did you have any difficulty deciding which questions were the most divergent and which were the most convergent? Did you try to answer these questions yourself? You might want to compare your answers with those of your classmates.

Questions for Evaluating

Another role of questioning in the learning process pertains to evaluation. In Part 4-1, you were introduced to the two functions of evaluation. Formative evaluation occurs *during* the learning process and is aimed at diagnosing and prescribing. In contrast, summative evaluation occurs *after* the instructional process and is used as a basis for assigning a grade.

Questioning is most frequently used by teachers in obtaining a summative evaluation. Generally, these questions tend to be content oriented, but this does not have to be the case, as you will see when you are introduced to the various questioning strategies. Summative evaluation questions are designed to find out how well the student has learned, and they are asked *after* the instruction is complete.

Formative evaluation questions, on the other hand, are asked at intervals *during* the instruction. They serve to keep the teacher informed of the students' progress. They also provide the students with immediate feedback. Adjustments in the instruction may result, or individual learner prescriptions may be given. There is no real structural difference between formative evaluation questions and summative evaluation questions. The difference is in the use or purpose of the question.

Bloom's Questioning Strategy

There are several well-known questioning strategies that offer guidance to teachers in planning and integrating effective questioning into the instructional sequence. Probably the most popular system used in classifying classroom questions is Bloom's *Taxonomy.*[1] An adapted version of the six levels or kinds of thinking identified by Bloom was presented in Part 4-1. The six levels are knowledge, comprehension, application, analysis, synthesis, evaluation. You may want to refer to page 195 to review these six categories.

These six sequential levels provide the teacher with a framework that can be used in formulating questions which will encourage the student to engage in a specific level of thinking in responding. The following outline presents each of the six levels of thinking as identified by Bloom. The intended student behavior for each level and some examples of questions which could elicit that intellectual behavior are also included.

Level One: Memory

The student is asked to recall information.

Examples of questions:

1. To which group of animals do frogs belong?
2. What are the four developmental stages of a moth?
3. What is the young undeveloped baby plant in a seed called?

Level Two: Comprehension

The student is asked to show understanding of the information.

Example of questions:

1. In your own words, how would you describe the four developmental stages of a moth?
2. Can you explain how fish are able to breathe under water?
3. We have been involved in gathering data concerning the amount of time a candle continues to burn when a beaker is inverted and placed over it. We have used three of four beakers and found that under beaker one (100 ml), the candle continued to burn for 11 seconds, under beaker two (200 ml), the candle burned for 22 seconds, and under beaker three (300 ml), it burned for 33 seconds. How long do you think it will burn under beaker four (400 ml)?

Level Three: Application

The student is asked to use abstract ideas and apply them to a specific concrete situation.

Examples of questions:

1. What would happen if you placed a fish in a covered container of cooled, boiled water?
2. How would you prepare an environment to grow frogs?
3. How would you use a balloon to demonstrate how our lungs work?

---

[1]Benjamin S. Bloom et al., *Taxonomy of Educational Objectives, Handbook 1: Cognitive Domain* (New York: David McKay Co., 1956).

Level Four: Analysis

The student is asked to examine information by separating it into its parts.

Examples of questions:

1. Why do you think people are not more concerned about pollution problems?
2. Timmy, can you explain how Robbie's conclusion is consistent with yours?
3. Brent, why is Courtney's approach sound?

Level Five: Synthesis

The student is asked to engage in creative thinking.

Examples of questions:

1. What could you do to find out how much water the bucket of snow would make?
2. How would you improve the pencil?
3. How would you describe life on Earth in the year 2077?

Level Six: Evaluation

The student is asked to make a judgment.

Examples of questions:

1. What is the best way to find out if water is a good conductor of heat?
2. Do you think that there should be a law to limit the number of children a person could have? Why? Why not?
3. What do you see as the best solution to the energy problem?

Bloom's six levels do represent a hierarchy. The knowledge level is considered to be the lowest level (simplest) of thinking and the evaluation level is considered to be the highest level (most complex). Each higher level includes any lower level. Be aware that asking a question designed to elicit thinking at a specific level does *not* mean the learner will automatically respond at that level. Careful planning is necessary. Quite often it is necessary to guide the learner to higher-level thinking by developing a series of questions that gradually move from the knowledge level to the comprehension level and on up to the higher levels. Learners cannot be expected to respond with evaluation level thinking until they have had many experiences involving the lower levels.

The students' background experiences and developmental levels must be taken into consideration when you plan effective questioning. As a teacher, you must provide many opportunities for students to gather, organize, analyze, and evaluate information. These experiences provide the necessary background that students need to draw upon when responding to questions. Without these kinds of experiences, students are unable to operate at the higher level of thinking.

Furthermore, a student's developmental level is a determining factor of the level of thinking at which she is capable of responding. Generally, young children before age seven or eight are unable to operate above the first three levels of thinking on Bloom's hierarchy. But that does *not* mean a teacher should never ask a higher-level question to a child below the age of eight. It *does* mean that these children need many more concrete experiences and that questions should relate directly to these concrete experiences.

Taba's Questioning Strategy

Hilda Taba has devised a questioning strategy which involves learners in a step-by-step process aimed at encouraging effective thinking.[2] In the first step, the teacher uses an opening question which requires only low-level cognitive responses, thus allowing most learners to enter the discussion. To be effective, an opening question should involve general knowledge but permit a wide range of responses within specific parameters. A question such as, "What are the four basic parts that most plants have?" is considered a closed question and would be inappropriate as an opening question. It has a "right" answer and does not encourage other meaningful responses once the right answer has been given. Using the same general theme, a teacher might ask, "How do the roots from a carrot, turnip, and beet differ?" This question allows more students to enter the discussion at a relatively low level since there are several possible responses.

The second criterion of an effective opening question involves establishing a focus for the discussion. The teacher must decide beforehand what the discussion's purpose is to be and guide the students in that direction. Consequently, the opening question should set the general direction for the discussion. This does not mean the teacher should attempt to control the content of the students' responses. It does mean, however, that he guides the students to engage in certain thought processes. For example, if the purpose of the discussion is to compare and contrast the roots from different types of plants, then the teacher must guide the students in that direction. The question used to open the discussion must be worded so that it relates to the initial purpose of eliciting low-level, factual responses as well as to the overall purpose of using this needed data in the higher-level cognitive processes of inferring and generalizing.

In the second step of Taba's scheme, the teacher attempts to raise the students' thinking from that of low-level memory or factual knowledge to a higher level in which they are required to look for relationships among data. Such questions are known as *lifting questions* and should be designed to encourage students to respond with tentative opinions rather than more qualified ones. Questions that include phrases such as, "what seems to be," "what might be," and "what do you think," are suggested in promoting open, tentative responses.

The following are examples of lifting questions:

1. Ashley, what factors might account for the difference in the size of your bean plant and Patrick's bean plant?
2. Adrian, what seems to be the best environment for growing bread mold, based on your own findings?
3. Peggy, what do you think will happen when we drop the raisins into the liquid in this beaker?

Not all students will be able to initiate a response at this level; so the teacher must attempt to involve as many students as possible by including other questions that extend the original responses, such as asking for other examples. In this way, a student who cannot raise the level of the discussion with an original response can participate once the level has been established by offering responses and examples

---

[2]Hilda Taba et al., *A Teacher's Handbook to Elementary Social Studies,* 2d ed. (Reading, Mass.: Addison-Wesley Publishing Co., Inc., 1971), chapter six.

similar to those initiated by other students. Frequently, a learner will respond with a thought that is correct but incomplete. When this happens, the teacher should continue with that student in order to permit him to develop the idea, provided the rest of the class can be kept mentally involved also.

In the third step of Taba's strategy, the student is required to combine relationships and make generalizations about them. During this phase of the discussion, the student is guided into thinking and speaking in abstract terms. Supporting questions which require the learner to clarify, extend, and synthesize are used during this phrase. These questions should encourage the learners to (1) build on the ideas already presented, (2) use gathered data as the basis for statements made, and (3) establish a rationale for their opinions. Questions such as, "Cheri, what evidence have we gathered that would support our hypothesis that gases expand when heated?" "Can someone explain why the balloon in the ice cold water changed shape?" or "What is the purpose of leaving one of the three balloons at room temperature?" permit learners to infer and generalize, thus operating at a higher cognitive level.

Taba also suggests that the teacher may find it helpful to let the students in on this step-by-step process of becoming a more effective thinker. The students can then be more directly involved in assessing their ability and their classmates' ability to respond at a higher level to different questions. Once the students become skilled at identifying the levels of thinking that different types of questions require, they are ready to assume more responsibility in the role of questioner. In the beginning, the teacher must assume the responsibility for designing and asking questions that require the students to respond at the different levels. As time passes, hopefully, the students will incorporate the process inherent in this strategy and begin to use it on their own in questioning the teacher, each other, and their textbooks.

The Taba strategy uses questions to direct the students' thinking step-by-step from low-level, specific concrete ideas toward more abstract, generalized concepts. It is suggested that the nature of the class as well as the purpose of the discussion are both important factors and must be considered when planning questioning sequences. There will be times when the teacher must supply needed information or use

This teacher is effectively using Taba's questioning strategy.

examples to clarify the situation so that the students can deal effectively with it. However, as the students gain more experience, the teacher's role should become more indirect.

It is important for the teacher and the students to work together in creating a classroom climate that allows for differences of opinion. The teacher should take the lead in offering support to students whose ideas are rejected or proven incorrect because it is important that these learners feel comfortable and secure enough to re-enter the discussion. The students themselves must also assume this supportive role.

The questioning strategies of both Taba and Bloom provide the classroom teacher with helpful guidelines in formulating a workable plan for effective questioning. No one series of questions will work every time and with all learners. It is up to the teacher to determine the specific purpose of a dialogue and survey the students who will be participating in order to devise effective questioning strategies. In addition, there are several other factors to consider. These include wait-time, verbal and nonverbal congruence, and the avoidance of teleological and anthropomorphic questions.

Importance of Wait-Time

Mary Budd Rowe has concluded that there is probably a relationship between the quality of inquiry and the wait-time a teacher allows for a learner to begin a response to a question.[3] This also holds true for the time allowed before the teacher replies to a student's statement. Many small groups of students and their teachers were involved in a study which explored the effects of increased wait-time. Before the study, the average wait-time allowed by the teachers was between one and two seconds for beginning a response and less than one second before replying to a student's statement. In the study, the teachers increased the wait-time in both instances and found that several changes occurred which led to an increase in the quality of the dialogue.

Not only were the student responses longer than before but also they initiated more of them. More learners were able to answer more of the questions. Relationships between evidence and inferences were better established, and more alternative explanations were offered by the students. It seems clear that increased wait-time on the part of the teacher can lead to a more effective discussion and, therefore, should be part of the overall questioning strategy.

Congruence of Verbal and Nonverbal Cues

Another factor that needs to be considered in implementing a questioning strategy is that of verbal and nonverbal congruence. In facilitating a dialogue, the teacher should pay particular attention to the nonverbal cues that the learners are receiving. The learners need to feel that the teacher is interested and is listening to what they are saying. Facing the child who is speaking and maintaining eye contact is one way to let the child know that you are interested in what is being said. Nodding appropriately, making facial gestures such as smiling, and making other gestures with the hands and body to indicate involvement with the discussion also demonstrate interest.

When the teacher is talking, the nonverbal and verbal cues need to be

---

[3]Mary Budd Rowe, *Teaching Science as Continuous Inquiry* (New York: McGraw-Hill, 1973), chapter 8.

coordinated so that students are not confused by words that indicate an involvement in the dialogue and body language that clearly indicates detachment. When conflicts between verbal and nonverbal cues occur, the nonverbal action tends to be taken by students as the more reflective of the two. Therefore, in order to facilitate an effective dialogue, the teacher must demonstrate both verbal and nonverbal involvement in the dialogue.

Avoidance of Teleological and Anthropomorphic Questions

Science questions which attribute purpose or will to nonhuman things and those that imply that natural phenomena have human characteristics should be avoided. These teleological and anthropomorphic questions are not consistent with scientific attitudes. When asking questions about natural phenomena, do not suggest or imply that nonhuman things think or feel in the same way that humans do. Questions such as, "How do you think a plant feels when it doesn't get enough water?" or "Why do some animals like to hibernate in the winter?" are misleading and do not help students understand natural phenomena.

Appropriate Response Schemes

As the facilitator of a dialogue, the teacher not only assumes the responsibility for designing and implementing the questioning sequence, but also must be aware of an appropriate response scheme. The purpose of a discussion should be to promote continued interaction among the participants. The teacher's response to a student's reply can impede the discussion or can serve to stimulate further thought.

The most common teacher responses are those that accept or reject the student's reply. These responses do nothing to stimulate further thought unless combined with a response that asks the learner to extend, clarify, support, or supply evidence. Continually giving accepting and rejecting kinds of responses is usually indicative of low-level questioning. Questions that require the learner to engage in high-level thinking call for teacher responses that encourage continued dialogue. Some examples follow of responses that could be used to stimulate dialogue:

1. Can you explain that further using your own words?
2. Can anyone else think of another example?
3. What evidence do you have to support your conclusions?
4. Adrian, can you add anything to Jeff's explanation?
5. Yes, Gladys, one function of the roots of a plant is the absorption of food materials. Can anyone tell us another?

Summary

Questions are considered a very important teaching tool. To be an effective, integral part of the instructional strategy, questioning sequences need to be carefully planned. Teachers need to be aware of the various instructional purposes for which questions can be used. Questions can be used successfully in any of the four phases of a lesson. Attention-getting and motivating questions can be used to begin a lesson. Questions can be structured to aid students in both the data-collecting and data-processing portions of the lesson. They can also be used to help students reach closure in the final phase of the lesson.

Moreover, questions can stimulate convergent or divergent thinking. Both kinds of thinking are necessary in most learning situations. The teacher must be able to determine which is most appropriate in a particular learning situation.

Information gathered by using evaluative questions can serve two functions: providing a basis for a grade and providing the teacher with information to be used in diagnosing and prescribing. If the purpose of the evaluation question is to elicit information to be used as a basis for assigning a grade, it is asked at the *end* of the instructional sequence. Evaluative questions asked *during* the instructional sequence can be used to keep the teacher informed of the learner's progress, and if warranted, adjustments can be made in the instruction or individual learner prescriptions given.

The next portion of the background information section described two well-known questioning strategies which offer guidance to teachers in planning and integrating effective questioning into the instructional sequence. Bloom has identified six sequential levels of thinking. He asserts that questions can be designed to elicit responses which involve the learner at each of the identified levels. These levels along with the intended student behavior and some examples of questions which could elicit that intellectual behavior were outlined in this section.

Hilda Taba's questioning strategy was also presented. This strategy begins with a broad opening question that encourages many learners to enter the discussion at a low cognitive level. It continues with a lifting question aimed at raising the learner's level of thinking. During this phase of the discussion, additional questions are asked that call for an extention of the original response in order to allow as many students as possible to continue in the discussion. In the final questioning phase, the student is guided into thinking and speaking in abstract terms. Questions which require the learner to clarify, synthesize, infer, and generalize are suggested. Taba's strategy also includes the goal of preparing the students to incorporate the process of the strategy and utilize it on their own when questioning the teacher, their classmates, and their textbooks.

Three other factors were discussed as having an effect on effective questioning. The wait-time a teacher allows for a learner to begin a response and also the time allowed before replying to the student's statement have been found to affect the quality of inquiry in discussions. The verbal and nonverbal cues from a teacher during a discussion need to be coordinated so that learners are not receiving conflicting signals. Many times, a teacher will be verbally involved in the dialogue, but her nonverbal cues, such as eye contact and facial gestures, indicate a detachment. Teleological and anthropomorphic questions should be avoided. These questions tend to imply or suggest that nonhuman things think and feel in the same way humans do. These types of questions are not consistent with a scientific attitude, are misleading, and do not help the child understand natural phenomena.

In addition to designing and implementing the questioning sequence, the teacher must also be aware of appropriate response schemes. In responding to a learner's reply, the teacher should try to stimulate further thought by asking the learners to extend, clarify, support, or supply evidence.

The activity section which follows has been designed to help you utilize the ideas presented in this background information section. You may work alone or with a small group in completing the activities. Some portions of the activities require interaction, and for these you will need to assemble a small group of your classmates in order to share information and ideas and receive feedback. You will want to work closely with your instructor throughout. There also are instructor seminars and discussion groups scheduled in the activity portion. You will probably want to refer to parts of the background information for guidance in completing the activities.

If you feel you need more clarification in understanding the background information, don't hesitate to ask your instructor for help. Perhaps you will want to schedule a small discussion group if others feel the need for further discussion. You might also want to explore some of the ideas presented in more depth. The bibliography and the suggested reading list provide excellent sources for a more in-depth look at many of the ideas and concepts presented. When you feel you are ready, go on to the activities.

As you complete these activities, keep in mind that they have been designed to help you assimilate and accommodate the information in the background reading section. You should try to analyze the purpose of each activity in order to relate it to the information presented. It might be helpful to reread the goals and objectives for this Part on page 228. Use them as a guide to what direction you should go in and what you are to achieve.

Activity 1: Formulating Science Questions

A. In this activity, you are to formulate several science questions designed to elicit higher than rote memory responses from elementary students. You may select a central theme, or concept, and relate all your questions to it, or you may choose to formulate questions from a variety of science topics. The important thing to remember is to design the questions so that the learners are encouraged to reply at a higher than rote memory level. You may work on these alone, or you may work with a small group.

Samples of science questions:

1.

2.

3.

4.

5.

6.

√ SELF-CHECK

1. If you did this activity alone, get together with several others and share your questions. You will probably need to refer to the background information in evaluating your efforts.
2. Use the following questions to analyze and evaluate your efforts:
   a. Do all your questions require more than a rote memory reply from the learner? Rewrite those that do not meet this criterion.

b. Do you or does anyone in your group have any questions that could be used effectively to (a) begin a lesson, (b) aid students in data collecting, (c) guide students in organizing and analyzing data, or (d) help students reach closure?

Try to think of other examples of questions that could be used for any of these purposes and discuss the possibilities with your group.

COMMENTS

c. Do you or does anyone in your group have any questions that are designed to stimulate convergent thinking? Divergent thinking? For what purpose would each kind be used? Refer to the questions on page 231 and discuss these with your group. Can you agree on which are most diverget and which are most convergent? Discuss any problems that occur. What other examples of divergent and convergent questions can your group think of?

COMMENTS

d. Do you or does anyone in your group have any questions that could be used in evaluating or assessing a student's ability? Could they be used just as effectively for formative as well as summative evaluation purposes? What other examples can you give of evaluative questions?

COMMENTS

e. Do you or does anyone in your group have questions which imply or suggest that nonliving things think or feel in the same way as humans? If so, change these so that the students will not be misled.

COMMENTS

Activity 2: Formulating Teacher Responses

A. In this activity, you are to work with a small group in formulating appropriate teacher responses to possible student replies. Select several of the questions from Activity 1 or formulate some new ones. Discuss the possible student replies to each of the selected questions. As each is discussed, formulate an appropriate teacher response. Record your responses so they can be discussed during the instructor seminar.

1. *Question:*

   Possible learner replies:

   Appropriate teacher responses:

2. *Question:*

   Possible learner replies:

   Appropriate teacher responses:

3. *Question:*

   Possible learner replies:

   Appropriate teacher responses:

√ SELF-CHECK

1. Do your questions generate several possible meaningful learner replies? If not, they probably encourage only low-level memory answers. Rewrite them, if necessary, so that they stimulate several higher than memory level responses.

2. Do your responses tend to stimulate further thought on the part of the students by encouraging them to extend, clarify, support, or supply evidence? If not, discuss how they could be altered to accomplish this purpose.

COMMENTS

Activity 3: Constructing a Questioning Strategy

A. In this activity, work alone or with a small group in selecting a science topic and constructing an appropriate questioning strategy as part of a planned science activity for elementary students. You may want to use the ideas and information presented in Unit 2, "Science Curricula Materials," in selecting your topic. Also, you may use the lesson, or portions of it, that you developed in Part 4-1, or you can develop another one. Your questioning strategy should incorporate the ideas presented in the background reading section of this Part. Use either of the two strategies presented, combinations of both, or formulate one of your own based on the information presented. Outline your strategy on a separate sheet of paper using the guide below. Give examples of questions that could be used. Limit your selected topic or content area so that it can be dealt with effectively in thirty to forty-five minutes.

Outline of Questioning Strategy

Science topic:

Grade level:

Brief description of lesson (purpose, content or concepts to be explored, and teaching strategy to be utilized):

Description of questioning strategy to be employed (include examples of questions):

B. Submit your questioning strategy to your instructor for feedback.

Summary: Questioning Techniques

There are various questioning strategies that offer guidance to elementary teachers in planning and integrating effective questions into the instructional strategy. In this Part, the themes and ideas of two of the more well-known questioning strategies were presented. Information about the purpose and kinds of questions was also included. You were made aware of factors to consider in designing your own questioning strategy for use with sciencing activities. Activity 3 involved you in constructing an appropriate questioning strategy as part of a planned science activity for elementary students. Review the goals and objectives listed on page 228. If you feel that you have met these goals and objectives, arrange for a final seminar with your instructor. If you do not feel that you have reached them, ask your instructor for individual help.

FINAL SEMINAR

Ask your instructor for directions in arranging for a final seminar to discuss this Part.

NOTES

COMPETENCY EVALUATION

Your instructor may choose to use a competency evaluation measure of some sort to evaluate your competency in the area of questioning techniques. See your instructor for specific directions.

BIBLIOGRAPHY

Bloom, Benjamin S. et al. *Taxonomy of Educational Objectives, Handbook I: Cognitive Domain.* New York: David McKay Co., 1956.

Bozardt, Delphine Anita. "Development of Systematic Questioning Skills in an Elementary Science Methods Course." Ph.D. dissertation, University of Georgia, 1973.

Carin, Arthur, and Sund, Robert. *Teaching Science through Discovery.* 3d ed. Columbus, Ohio: Charles E. Merrill Publishing Co., 1975. Chapter 6.

George, Kenneth. *Elementary School Science: Why and How.* Lexington, Mass.: D.C. Heath Co., 1974. Chapter 4.

Hunkins, Francis P. *Questioning Strategies and Techniques.* Boston, Mass.: Allyn & Bacon, 1972.

Riegel, Rodney, "Classifying Classroom Questions." *Journal of Teacher Education* 27, no. 2 (1976): 156-61.

Rowe, Mary Budd. *Teaching Science as Continuous Inquiry.* New York: McGraw-Hill, 1973. Chapter 8.

Taba, Hilda, Durkin, Mary; Fraenkel, Jack, and McNaughton, Anthony. *A Teacher's Handbook to Elementary Social Studies.* 2d ed. Reading, Mass.: Addison-Wesley Publishing Co., 1971. Chapter 6.

SUGGESTED READING

Carin, Arthur, and Sund, Robert. *Teaching Science Through Discovery.* 3d. ed. Columbus, Ohio: Charles E. Merrill Publishing Co., 1975. Chapter 6.

George, Kenneth. *Elementary School Science: Why and How.* Lexington, Mass.: D.C. Heath Co., 1974. Chapter 4.

Rowe, Mary Budd. *Teaching Science as Continuous Inquiry.* New York: McGraw-Hill, 1973. Chapter 8.

Taba, Hilda, Durkin, Mary, Fraenkel, Jack, and McNaughton, Anthony. *A Teacher's Handbook to Elementary Social Studies.* 2d ed. Reading, Mass: Addison-Wesley Publishing Co., 1971. Chapter 6.

UNIT 5
Classroom Management

The hands-on approach to science teaching demands that teachers be skillful in classroom management. Although a teacher's knowledge of and skill in teaching science are necessary, they alone will not produce a successful learning experience. Previously, you were made aware of some techniques to use in helping students acquire science knowledge, skills, and attitudes. In this Unit, you will explore two areas related to classroom management: physical environment and safety.

The physical environment of the science classroom has both a direct and indirect influence on the kind of learning that takes place. Four areas of concern related to the management of the physical environment (along with ideas and suggestions for controlling and directing it so that learning is enhanced) are explored.

Teachers must also plan for student safety. This involves being aware of potential hazards in any given activity or situation. Help in identifying potential hazards and suggestions for decreasing the likelihood of injury are explored in this Part.

In each of the identified areas, background information relative to the specific topic is given. You are encouraged to discuss the information and ideas presented with your classmates and your instructor. Activities are included which are designed to help you develop skill in classroom management.

PART 5-1

Physical Environment

FLOWCHART: Physical Environment

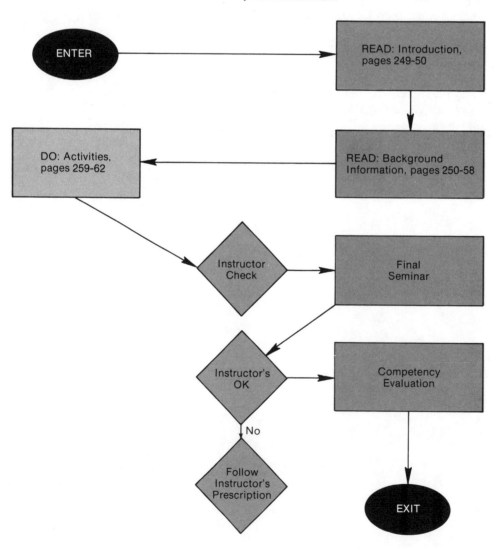

Introduction

As a teacher, you must become increasingly sensitive to the physical environment of the classroom and be able to control and direct it so that learning is enhanced. The physical environment has both a direct and indirect influence on the kind of learning that takes place. Many an otherwise well-planned lesson has failed because of the environment. There are four areas of concern that will be discussed briefly here:

1. Physical factors—lighting, room temperature, desk size, and distracting noises and sights
2. Classroom arrangements—physical furnishings and grouping of students
3. Organization and storage of materials and equipment
4. Special problems of students—physical limitations, cognitive limitations, social and/or cultural limitations

While reading and discussing the information included here, think of your own experiences in elementary school science. Think of specific examples when the physical environment enhanced or inhibited your learning experience. As you observe others in mini-teaching experiences or during student teaching or as you engage in the activities included here, begin to look for ways the physical environment can directly or indirectly affect the learning experience.

GOALS

After completing this Part, you will demonstrate competency in the following:

A. Identifying and controlling factors in the physical environment that could enhance or inhibit learning experiences
B. Constructing classroom arrangements that would facilitate learning
C. Organizing materials and equipment used in science activities
D. Recognizing and providing for special problems of students

BEHAVIORAL OBJECTIVES

In completing this Part, you will do the following:

A. Identify and describe means of controlling the following physical factors in order to enhance learning:
    1. Classroom lighting
    2. Room temperature
    3. Desk assignment
    4. Distracting noises and sights
B. Identify and construct classroom arrangements that are appropriate for the following purposes:
    1. Large group instruction
    2. Small group instruction
    3. Individual instruction

C. Identify and explain how the following techniques can be used in organizing and utilizing science materials and equipment:
   1. Labeling
   2. Using tote trays, dishpans, buckets, and boxes
   3. Posting a materials list on cabinets and drawers
   4. Making a master list of equipment and material location
   5. Providing proper storage containers
   6. Providing appropriate storage arrangement
D. Identify the following as special problems students might have and describe appropriate ways of providing for them:
   1. Physical limitations
   2. Cognitive limitations
   3. Social and/or cultural limitations

Background Information

Physical Factors

A most important factor to consider when checking a classroom's physical environment is lighting. Improper and inappropriate lighting in a classroom can inhibit learning. Both natural lighting and artificial lighting need to be surveyed to determine possible problems. Look for glare in the room, particularly on the chalkboard. Shadows can also be a problem. Check the various work areas of the room to see that light is adequate there. Make your checks at different times of the day and during different weather conditions to account for the influence of natural light. Once the problem areas are identified, possible solutions can be generated and put into effect. These solutions may result in structural changes in the classroom itself or in changes in the type and/or arrangement of the lighting fixtures. Such solutions are usually long range and expensive. Less expensive solutions can be utilized. Shades, blinds, curtains, cardboard, or other opaque and translucent objects and materials can be used to effectively control the amount and direction of both inside and outside light. If the specific areas within the classroom need more direct light, extra light sources, such as lamps and flashlights, can be made available. Work areas needing less light, such as those set up for filmstrip viewing and using the opague projector, can be located in the darker areas of the room. Bookcases or cardboard walls can be used to block out some of the unwanted light in these areas.

Room temperature is another physical factor that can impede learning. When classrooms are too warm, children become drowsy and listless. On the other hand, cold classroom can cause the learners to become preoccupied with their need for physical comfort. Thus, physical discomfort can cause a lack of attention and result in a poor learning experience. If the classroom temperature cannot be directly controlled by adjusting the thermostat, other measures must be taken. If the room is too warm, fresh air can be circulated by opening a window or door. A fan can also be used to help circulate the room air. Shades and heavy curtains or outside awnings can be used to block some of the direct sun rays that can cause a room to become overheated. Furthermore, if the room temperature changes frequently, the children can be instructed to wear "layered clothing" that can be added to or taken off as the temperature fluctuates.

One important factor to consider in the physical environment of the classroom is desk size.

Desk size is another important consideration. Children are all sizes, and they need appropriate desks. Therefore, various desk sizes need to be available in the classroom. Feet should be able to rest comfortably on the floor. The desk top should be at a level that is comfortable for writing. A child's knees should not touch the writing desk. Generally, flat, table-like desks are best. They can be pushed together to form larger working areas for group projects.

Distracting background noises and sights are factors of the physical environment that can cause children's attention to be drawn away from the planned learning experience. Minimize these by having children face away from the possible distraction. For example, don't let a caged animal or a window with a view of the playground be your backdrop. Sometimes quiet background music helps to minimize other distracting sounds.

As a teacher, you will need to be alert to the many factors in the physical environment that could interfere with the science experience. Some of the more common ones have been discussed here. However, as you gain experience in the classroom, you may identify others. Many times, the problem factor cannot be totally eliminated, but it can be minimized by utilizing alternative solutions. Perhaps you will want to involve your students in the process. After all, identifying problems and generating alternative solutions are important aspects of sciencing.

Classroom Arrangements

Both the arrangement of classroom furnishings and the grouping patterns of the students have an effect on the learning process. This is true for the self-contained classroom—one grade, all subjects—as well as for the departmentalized classroom —just science for one or more grades. The same general problems exist in both kinds of classrooms. First, let's focus on the area of classroom furnishings.

As a teacher, you will probably have little input into the design of your classroom. Most likely, you will simply be assigned a classroom and be expected to utilize the already existing furnishings. Given this probable situation, let's explore some of the possibilities open to you. Your classroom will contain both *permanently placed* and *movable* furnishings. Permanently placed furnishings, such as the sink, electrical outlets, windows, chalkboard, lighting fixtures, and perhaps a closet area, must become the core of your classroom arrangement. The movable furnishings, such as desks, tables, bookcases, plants, and an aquarium, can be used to modify and enhance a core arrangement.

No *one* room arrangement will be appropriate for every science experience. But, generally there are four designated areas that should be considered in any arrangement. One area of the classroom should be arranged to accommodate the manipulative aspect of sciencing. In this area, students find the materials, equipment, and facilities necessary for firsthand data gathering. Another part of the classroom should be designated as a resource center. Here, books, magazines, newspapers, cassette tapes, records, filmstrips, film loops, pictures, models, and other similar sources of information should be available for student use.

This area has been set aside for firsthand data gathering.

A small area that can be used for individual conferences or small group discussions needs to be located away from the center of activity. A certain amount of privacy should be provided in this area so that students who wish can talk freely with the teacher. This area can also be used for individual testing and tutorial sessions. Storage areas should also be a part of the overall room arrangement. Some areas will be needed to house the materials and equipment in use, while other areas must be used to store supplies not in immediate use. Facilities must accommodate both large and small items. Living things, such as small animals, fish, plants, and insects, require special attention. More specific ideas for utilizing storage areas are explored in the next section, "Organization and Storage of Materials and Equipment."

The way the furnishings in a classroom are arranged can definitely affect learning outcome. Therefore, teachers must provide arrangements that facilitate learning. Students should have easy access to the various materials, equipment, and facilities to be used in a learning experience.

The second area of concern in classroom arrangements is the way in which students group themselves or are grouped by the teacher. Much sciencing is done in small groups, partly to accommodate for the lack of equipment and materials in most elementary classrooms. However, even when there is an abundance of science materials, students need to have the experience of working, sharing, and communicating with their peers. The process aspect of science is enhanced at the elementary school level by increased interaction among students.

Various types of groupings can be used to facilitate learning. Three grouping patterns will be discussed here:

1. Individual grouping
2. Small groups (two-five people)
3. Large groups (six or more people)

Some science experiences are best accomplished when each student works independently. Upper elementary students can be expected to work alone on projects much more successfully than can younger students. Highly motivated students may prefer working alone much of the time. Both the age of the student and individual motivation should be considered in grouping students for science activities. Each specific science experience should also be considered in order to determine the appropriate type of grouping. Some experiences can be done very successfully alone while others require constant interaction among the students. Different grouping patterns may also be used during one activity. For example, during the data-gathering phase, students may work alone, but they may find it helpful to interact with others during the data-processing phase.

As mentioned earlier, the small group seems to work well for many types of science experiences. Three ways to organize small groups for science experiences are suggested here. Sometimes students should be allowed to determine with whom they would like to work. They may form their small group on the basis of friendship, common interest, or some other factor of their own choosing.

Other times, the grouping should be done by the teacher. For example, students could be grouped to minimize behavioral problems, to take advantage of their special skills, or to encourage verbal participation and leadership.

Random assignment can also be used effectively in grouping students. Occasionally, students enjoy being placed in a group on the basis of luck. For example, numbers could be drawn from a hat to form the grouping. Often, in this manner, relationships can develop between students who otherwise might never have worked together.

Large group instruction is most successfully used for giving information and directions and for summarizing. During large group instruction, students are generally put in a passive role. Interaction becomes limited; otherwise, chaos can easily result. Large groups serve quite well for purposes, such as films, some demonstrations, and lectures.

Grouping should be done purposely to facilitate learning. Teachers must determine which type of grouping is appropriate for each activity.

Organization and Storage of Materials and Equipment

Poor organization of science materials and equipment often is the major cause of an unsuccessful science program. Most teachers would "do science" more often if materials and equipment were arranged for easy access. Some practical ideas for storing science equipment and materials and suggestions for organizing in-use materials are explored here.

A master list of on-hand materials and equipment is invaluable. The list can be alphabetized or items can be grouped according to identified headings. The exact location of each item should be included. This list can be organized to correlate with specific lessons or broad topic areas. Cabinets, drawers, shelves, boxes, and other storage facilities being used should have a materials list posted which identifies the items that can be found there.

Keeping materials organized is important to a successful science program.

All items should be labeled to facilitate easy identification. Harmful chemicals and other hazardous materials must be so labeled. A designated area might be used to house "off limits" materials and equipment. These types of materials should be kept in a rather isolated area of the room in facilities that can be locked. Appropriate containers, preferably plastic, should also be provided. More detailed suggestions regarding safety in the classroom are provided in Part 5-2, "Safety."

Teachers have used many techniques for organizing needed materials and equipment. Shoe boxes are just the right size for storing materials and small equipment for one or two persons. They stack well and can be acquired easily. Plastic dishpans and buckets serve well for holding materials for small groups. They can easily be carried from a storage shelf to a table where several students could share the materials. You can probably think of other similar ways to make materials easily accessible to elementary students.

This teacher has organized each activity into individual plastic bins.

Frequent inventories will help to keep the materials and equipment organized for ready use. Consumable items must be replaced as they are used, and equipment must be kept in good repair. A teacher will want to make sure that sufficient amounts of needed materials are always available.

An integral part of most science lessons is the distribution of materials at the beginning of the lesson and the collection of them when the lesson is finished. Both of these procedures can be accomplished with ease by planning appropriately. A simple routine can be established using "helpers" who pass out the equipment and materials. If boxes, buckets, or dishpans containing only the needed items are used, this task becomes simple. Children should be directed to clean all items after using tnem and return them to the appropriate box, bucket, or dishpan. Consumable materials must be replaced, and all boxes, buckets, and dishpans returned to the storage areas.

To avoid confusion, make sure that elementary students are able to operate any equipment to be used in a science experience. Frequently, this may mean spending one or more days just learning how to manipulate the equipment before actually beginning the lesson. Written directions should be provided for each student. These should be given orally as well. Generally, short steps are more easily followed than long, detailed directions.

Special Problems of Students

Good classroom teachers have always been responsive to the individual needs of the children they teach. With the growing trend of mainstreaming students with special problems into the regular classroom, the task of meeting individual needs has become much more diversified. Most elementary curriculum materials developed before 1975 do not include specific suggestions for the classroom teacher to use in adapting the program to fit the needs of children with special problems. The purpose of this section is twofold: (1) to help you build an awareness of the special problems children might have which could affect learning and (2) to encourage you in developing and adapting science activities and equipment for use by children with special problems.

Children's learning can be affected by the special problems that result from the following:

1. Physical limitations
2. Cognitive limitations
3. Social and/or cultural limitations

Teachers must begin to develop means of providing appropriate learning experiences for children with these special problems, within the framework of the regular classroom. Individualization must be taken a step further to include *all* children. Each of these identified areas will be explored briefly in order to increase your general awareness.

PHYSICAL LIMITATIONS

Physical limitations can take various forms, all of which can interfere to varying degrees with learning. Children who have limited or impaired sight and hearing will be unable to participate in most science activities unless special provisions are made. These children must be provided with other means of accomplishing a task that would ordinarily be dependent upon sight or hearing. In most instances, this can be accomplished by modifying the activity so that visual information, for the blind child, or information gathered through hearing, for the deaf child, is gained through the other senses.

This approach is being used with a group of blind, deaf, and emotionally disturbed children who are mainstreamed into regular classrooms at the Horace Mann School in the District of Columbia.[1] In order to accommodate for the blind and deaf children in the regular classroom, the science teacher uses both visual cues and language in conveying information to the students. In some activities, additional equipment is provided for the special students. For example, in an experiment involving the factors affecting the swinging of a pendulum, a light sensor which emits a beeping sound is used to help a blind student detect the rate of the swinging pendulum. As the pendulum swings, it cuts through the light wave, causing a change in the volume and pitch of the beeping. By listening for this change, the blind student can gather the needed information.

Science Curriculum Improvement Study (SCIS) has available braille versions of the student manuals. They also have special teacher helps available which offer suggestions for working with visually impaired children. Regular classroom teachers must be aware of the special problems caused by impaired vision and hearing and must provide ways for obtaining information that utilizes a child's other senses.

Children with limited mobility must also be considered in planning science activities. Some modification of the activity may be necessary or alternate activities provided. In some cases, children can work together, each using their own special abilities in completing an activity.

Allergies can also be considered a physical limitation. Some children will be unable to participate in science activities that involve live animals or certain plants without an allergic reaction.

---

[1]Efthalia Walsh, "Lab Classroom: Breaking the Communication Barrier," *Science* 196 (June 1977).

Children's physical limitations can affect their learning. Although there are some science materials available for use with children with physical limitations, these are limited. Most likely, the classroom teacher will be responsible for developing and adapting existing materials for use with children who have such limitations.

COGNITIVE LIMITATIONS

Cognitive limitations affect children's learning in a different way than do physical limitations. Therefore, a teacher must make different kinds of modifications in the science program to fit the needs of children with cognitive limitations than those made for children with physical limitations. For children with cognitive limitations, science concepts must be presented in a concrete manner. Prolonged manipulation of equipment and materials may be necessary. Activities should center around one aspect of a concept at a time rather than many. Constant verbalization of actions and thought may be necessary to help children understand a concept. The time spent on any one activity will need to be adjusted to the attention span of the child. Most children with cognitive limitations are easily distracted; therefore, only those materials needed for a specific activity should be available. *Elementary Science Study* (ESS) has assembled a *Special Education Teacher's Guide.*[2] This guide is designed to help special education teachers utilize the regular ESS materials with special children. Its suggestions and ideas can also be used effectively by regular classroom teachers. Even though the guide is specifically for use with the ESS program, there are many general ideas and suggestions that are applicable to any science program. Children with cognitive limitations can have successful science learning experiences. Teachers must be aware of the nature of the limitation and must provide appropriate instruction.

SOCIAL AND/OR CULTURAL LIMITATIONS

Social and/or cultural limitations can also inhibit a child's learning. This is particularly true if a child's home environment is vastly different from the one at school. Teachers must provide instruction that will help bridge the gap between the two. Science activities can be modified so that a child's immediate environment and past experiences are used in spanning the gap. This allows the child to relate more directly to a science concept or process, thus resulting in a more meaningful science experience. Once a link between a science concept or process and the child's world has been established, it is possible to build upon this link and the child's experiences.

In the activities, you will examine some materials that have been adapted for use by children with special problems. You will also work with a small group of your classmates in generating ideas for adapting science activities and equipment for use by children with such problems. A list of suggested reading is provided at the end of this Part for those who would like more specific information and direction in providing appropriate learning experiences for children with special problems.

---

[2]Daniel W. Ball, *ESS/Special Education Teacher's Guide* (St. Louis: Webster/McGraw-Hill, 1978).

Summary

Four areas of concern related to the physical environment of the science classroom were explored here. Such things as lighting, room temperature, desk size, and distracting noises and sights were identified and discussed as physical factors that can enhance or inhibit learning. As a teacher, you will need to be alert to these and other such factors in the physical environment that could interfere with an otherwise well-planned science experience.

The arrangement of classroom furnishings and the grouping patterns of students are other areas of concern that were explored. Suggestions for arranging the classroom to facilitate the various aspects of science—manipulation of materials, use of resource material, individual or small group conferences, and storage of materials—were provided. Three ways of grouping students to facilitate learning were also explored.

A third area discussed was the organization of science equipment and materials. Several techniques for appropriate storage and easy retrieval of both in-use and on-hand materials were suggested. Ideas related to the distribution and collection of materials were also discussed.

The final area of concern focused on the special problems some students might have that could interfere with learning. Three problem areas were discussed briefly: physical limitations, cognitive limitations, and social and/or cultural limitations. It was suggested that a teacher must be sensitive to the special problems a student might have and be prepared to modify existing procedures or develop new ones in providing meaningful learning experiences for *all* children.

The following activities have been designed (1) to help you become more sensitive to the physical environment of the classroom and (2) to help you develop skills in controlling and directing the physical environment so that learning is enhanced. Four clusters of activities are provided. Each cluster relates to one of the four areas of concern identified in the background information. It is hoped that by participating in these activities, you will be able to utilize some of the ideas and suggestions from the reading section.

Activity 1: Controlling Physical Factors

A. Using the room in which your class is held, work with a small group in identifying and suggesting ways of controlling the following physical factors in order to enhance learning:

1. Classroom lighting
2. Room temperature
3. Desk assignment
4. Distracting noises and sights

Organize and record your group's recommendations.

COMMENTS

B. Arrange to visit an elementary classroom during a science lesson. Observe how the lighting, room temperature, desk assignment, and distracting sights and sounds affect the learning experience. Record your observations and offer suggestions for improvement. Be sure to make note of the ways in which each of the above-mentioned factors are used to enhance the learning experience. Share your findings with a group of your classmates.

OBSERVATIONS

RECOMMENDATIONS

COMMENTS

Activity 2: Arranging Classrooms

A. On a separate sheet of paper, identify some common classroom furnishings, such as desks, tables, bookcases, chalkboard, sink, etc. Draw separate diagrams illustrating how the identified furnishings could be arranged for each of the following purposes:

1. Large group instruction
2. Small group instruction
3. Individual instruction

Remember, this is the *same* room with the *same* furnishings, but they are arranged in different ways in order to facilitate the type of specified instruction.

B. Share and discuss your arrangements with a small group of your classmates. Make note of any new ideas you get from this group sharing.

Note: This activity can be modified so that a real elementary classroom is used. The actual furnishings can be listed and the room drawn to scale, or if arrangements can be made, the actual manipulation of the furnishings can take place.

Activity 3: Organizing Equipment and Materials

A. Work with a small group in checking the room in which your class is held in order to identify various techniques that are utilized in organizing science equipment and materials. List these.

B. Discuss your findings with your group. What recommendations can your group make that would improve the existing organization of the science materials? List your recommendations.

C. As a group, choose one of the recommendations you listed and make plans for implementing it. Discuss your plans with your instructor for approval, then put them into action.

Note: This activity can be done using an elementary classroom if one is available. You might want to discuss this possibility with your instructor.

Activity 4: Adapting Instruction for Special Problems

A. Examine all of the materials your instructor has on hand related to providing appropriate instruction for children with the following special problems:

1. Physical limitations
2. Cognitive limitations
3. Social and/or cultural limitations

B. Review several of the recent educational journals and magazines for information and ideas regarding instruction for special children.

C. Form a small group and select a science activity designed for the regular classroom from a textbook or science program. Working with your group, modify the activity so that children with special problems could participate. Write up your ideas on a separate sheet of paper and submit them to your instructor for feedback.

BIBLIOGRAPHY

Ball, Daniel W. *ESS/Special Education Teacher's Guide.* St. Louis: Webster/McGraw-Hill, 1978.

Carin, Arthur A., and Sund, Robert B. *Teaching Science through Discovery.* 3d ed. Columbus, Ohio: Charles E. Merrill Publishing Co., 1975. Chapter 12.

Cooper, Katherine E., and Thier, Herbert D. "Do You Have to See It? Laboratory Science for Visually Impaired Children." *Learning,* April 1974, pp. 44-45.

Rowe, Mary Budd. *Teaching Science as Continuous Inquiry.* New York: McGraw-Hill, 1973. Chapters 13, 14, 15.

Thier, Herbert D., and Hadray, Doris E. "We Can Do It Too." *Science and Children,* December 1973, pp. 7-9.

Walsh, Efthalia. "Lab Classroom: Breaking the Communication Barrier." *Science* 196 (June 1977).

SUGGESTED READING

Carin, Arthur A., and Sund, Robert B. *Teaching Science through Discovery.* 3d ed. Columbus, Ohio: Charles E. Merrill Publishing Co., 1975. Chapter 12.

Cooper, Katherine E., and Thier, Herbert E. "Do You Have to See It? Laboratory Science for Visually Impaired Children." *Learning,* April 1974, pp. 44-45.

Rowe, Mary Budd. *Teaching Science as Continuous Inquiry.* New York: McGraw-Hill, 1973. Chapters 13, 14, 15.

Thier, Herbert D., and Hardray, Doris E. "We Can Do It Too." *Science and Children* December 1973, pp. 7-9.

Summary: Physical Environment

You have completed all of the instruction included in this portion of Unit 5. Refer to the behavioral objectives on pages 249-50. You may find it helpful to skim through the background information again and the activities. If you feel that you have *not* reached competency in this area, ask your instructor for individual help. If you feel you are able to demonstrate the behavioral objectives, arrange for a final seminar with your instructor.

FINAL SEMINAR

Get together with a small group of your classmates to discuss this Part. *Invite your instructor to join you* for this final seminar.

NOTES

COMPETENCY EVALUATION

Your instructor may choose to use a competency evaluation measure of some sort to evaluate your competency in the area of classroom environment. See your instructor for specific directions.

PART 5-2

Safety

FLOWCHART: Safety

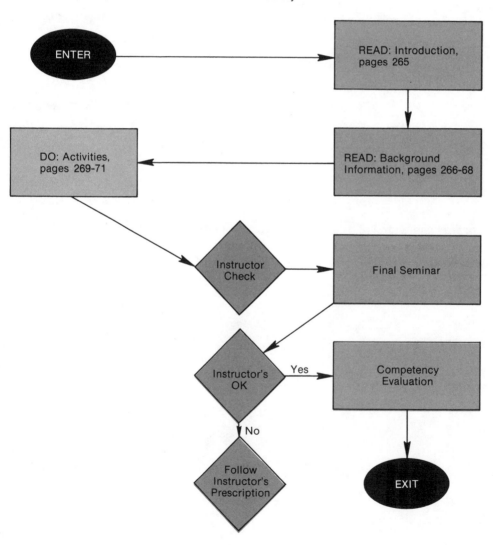

Introduction

Elementary school teachers have long been responsible for the safety and well-being of students entrusted to their care. Perhaps because this concept has become so much a part of a teacher's daily responsibilities it is sometimes taken for granted. Consequently, care and conscientious attention to detail are more and more frequently omitted.

Teachers must plan for student safety. Such planning requires foresight and insight. Teachers must be aware of the potential hazards in any given activity or situation. Moreover, care must be exercised in planning lessons to plan for, or even better to eliminate, situations in which an injury might occur.

Teachers are not born with skill in foreseeing hazards any more than they are born with skill in planning lessons. Part 5-2 will help you in developing these needed skills.

GOALS

After completing this Part, you will demonstrate competency in the following:
A. Identifying safety hazards in the science classroom
B. Utilizing techniques aimed at prevention of injury to students
C. Understanding tort liability in science teaching

BEHAVIORAL OBJECTIVES

In completing this Part, you will do the following:
A. Identify potential safety hazards in a classroom situation
B. Suggest ways to prevent injury to students in a classroom situation which involve:
    1. Reducing hazards and/or
    2. Utilizing alternate methods
C. Locate, examine, and operate selected safety equipment
D. Collect and analyze information provided by the State Department of Education concerning O.S.H.A. (Occupational Safety and Health Act)
E. Explain tort liability by listing and defining each of the necessary components
F. Identify the following as being of particular concern to science teachers in avoiding injury and tort liability:
    1. Careful planning
    2. Storing and labeling hazardous substances in the classroom
    3. Giving explicit directions each and every time an activity is conducted
    4. Supervising students at all times

Part 5-2 was co-authored by Dr. Anita Bozardt, Oakland University.

Background Information

By attempting to identify hazardous situations, teachers become more aware of such situations. Certainly every teacher would caution students when boiling water on the stove; perhaps students would not be allowed to do it at all. Some things are fairly obvious; we're all aware that such a situation poses a potential hazard: the student might be burned. However, do we all recognize the potential hazards in a simple activity such as the following.

A kindergarten class has been developing observation and simple classification skills. The teacher has planned a lesson using beans, peas, marbles, and other small items. During the lesson the teacher is to give the students containers of objects to be sorted into three groups according to some criteria. The students are to put the objects into three different baby-food jars to demonstrate their skills of observing and classifying. Can you identify at least five potential hazards in this situation?

1.

2.

3.

4.

5.

Did you think of these?

1. Having allergic reactions to handling the objects
2. Throwing the objects and perhaps injuring another student
3. Putting the objects in ears, nose, or mouth and having them become lodged there
4. Breaking the jars and sustaining a cut
5. Spilling the objects with the result that someone steps on one, falls, and becomes injured

Did you think of something not listed here? If so, please share it with your instructor and the class.

Students are encouraged to do activities in the newer science curricula. This hands-on approach requires that they manipulate materials and equipment frequently. Consequently, potential hazards abound in a "with-it" classroom.

To decrease the likelihood of injury, teachers could eliminate the manipulation of materials. However, that would defeat our purposes in science education. More logically, plans could be made in such a way as to identify hazards in advance. Alternative methods to accomplish the same goal, without the hazard, could be substituted. If that were impossible, at least the students could be warned, and techniques could be demonstrated for using extreme caution.

Students should be taught proper techniques in handling and caring for materials and equipment. This step is more likely to occur when expensive materials or equipment are being used (for example, the expensive microscope or rare book). However, the cost of an item does not indicate its relative potential as a hazard. Even "collectables" can be handled in such a way as to result in injury. *Proper techniques can and should be taught.*

Most teachers make rules for proper conduct in the classroom and designate specific procedures to be followed in certain situations. Likewise, rules should be made concerning the use and handling of science materials and equipment. Perhaps a specific area of the room can be designated as the science area. Rules concerning the use of selected equipment (the hot plate maybe) might specify that the equipment will be used only when the teacher is assisting the student. Reminders can be posted in the form of charts or cartoon-like posters.

In the activity example provided above, one danger is spilling the objects, stepping on one, and falling. The danger might be reduced somewhat by limiting the area in which the activity is conducted. Perhaps a carpeted area might reduce injury from a fall and the chance of breaking the jars if they were dropped. In planning that lesson, the teacher could plan to carry it out in a carpeted area of the room. In addition, the teacher could avoid part of the potential danger by using nonglass containers.

Safety guides have been developed in many states and local school districts. If one is available in your state or area, obtain a copy and read it carefully. Then save it for future reference. The hints and suggestions in such guides can save a lot of time and perhaps even save a child from injury.

A technique of utmost importance in avoiding pupil injury is to give clear, complete directions each and every time an activity is to be performed. Do not assume your students will remember even general directions or rules from one day to the next or from one manipulation to the next. Give directions and repeat rules each time.

With the growing practice in our society of people entering a lawsuit against someone for minor, or maybe even imagined, injuries or slights, teachers need to develop an awareness of such action and protect themselves against it. Pupils can and do bring lawsuits against teachers for injury. The best defense is to make sure that no students sustain injury in your class. However, no one can absolutely insure against injury. A student will occasionally become injured even with the most careful planning. In that case, a teacher's best legal defense is the careful planning. Without

This teacher stresses safety guidelines.

careful planning, a teacher could be found negligent and then perhaps legally liable for the student's injury. Teachers must guard against injury and against negligence.

A standard technique for determining possible negligence is to ask the following question, Did the teacher do everything any reasonable person would do in a similar situation? Teachers should keep this question in mind and ask it of themselves. At this phase of your professional development, it would be a good idea to establish the habit of asking this question as a self-check. Ask it of yourself each time you write a lesson plan.

It is not the purpose of this section to scare propsective teachers, only to build awareness. Prevention is so much better than regret. Therefore, plan lessons carefully; think about safety needs; caution students; and keep hazards to a minimum. Ask yourself, "Have I done everything any reasonable teacher would do in a similar situation?" Do not become careless or be lulled into a sense of false security.

Some teachers believe that the school board or the state would be legally liable for a student's injury. True, teachers do work for the state. However, many states adhere to the doctrine of sovereign immunity, which means that one cannot sue the state in cases of injury resulting from negligence of an employee. In such states, the teacher, individually, would face possible conviction in a tort liability suit.[1] (Of course in all states the individual teacher is responsible for criminal acts.)

Summary

The topic of tort liability is far too complex to cover completely here. The intention here is to build awareness. Awareness can help teachers avoid situations which would leave students exposed to possible injury and themselves vulnerable to a claim of negligence. Careful planning is the best prevention. Try at all times to keep your students safe from injury. Many safeguards are simple, common-sense precautions:

1. Make and enforce rules
2. Label dangerous substances and store them out of the reach of children
3. Avoid hazardous situations
4. Remind students regularly of safety rules
5. Supervise students at all times

You will be able to do these things. The problem is remembering to do them. You must remember, for the children's safety and for your safety.

---

[1]Tort liability is liability for an injury to an individual. Civil liability refers to wrongs committed against society.

Activity 1: Locating Emergency Areas

A. Work with a small group to check the building in which this class is held for location of the following items:

1. Exits
2. Fire extinguishers or hoses
3. Fire alarms
4. Safety showers
5. First aid kits
6. Fire escape routes
7. Storm or air raid shelters
8. Running water sources

B. Make a list or map and check yours with another group. Do this activity again in the school to which you have been assigned for your field experience. Discuss your findings with your instructor.

COMMENTS

Activity 2: Operating a Fire Extinguisher

A. Obtain a fire extinguisher from your instructor. Read the directions carefully. In an area specified by your instructor, operate the extinguisher as if you were actually going to extinguish a fire. Learn the differences in types of fire extinguishers. Be sure you have actually used one. That will save time if you ever need to use one in an emergency.

COMMENTS

Activity 3: Identifying Potential Hazards

A. Do one of the following with a small group of your classmates:

1. From an elementary school science textbook, select two chapters and identify all possible hazards for a class doing the activities.
2. Using lesson plans you have written for this class, select four and identify all potential hazards.

3. For a situation provided by your instructor, identify all possible hazards.
4. Observe an elementary school class for two hours (including the science period, if possible). Identify and list all potential hazards to student safety. Record your responses.

B. For two of the hazardous situations identified in A of this activity, suggest ways to prevent injury to students. How could the hazard be removed or reduced? Give specific suggestions. What alternative methods could be used in the lesson? Record your responses.

C. Share your ideas with your instructor.

COMMENTS

Activity 4: Recognizing Safety Rules

A. Ask the principal of an elementary school about the safety rules of that school or district. Is there a safety guide for the teachers? Obtain a copy of the list and/or guide and share it with your instructor. Save the list for the next part of this activity. Does the list have a rule about using open flames or candles? Are goggles required? What special requirements are there for field trips? Is there a rule about washing hands? When do pupils have to wear an apron or other clothes protector?

B. Do either of the following:

1. Make pictorial posters as reminders of safety rules for use in your classroom later. (Posters may be 8½″ x 11″ or larger.)
2. Make and display a bulletin board for safety. Include safety rules listed in A of this activity. Maybe you can try your posters or bulletin board in an elementary school classroom. The children can provide helpful feedback and creative suggestions. Arrange this with your instructor if you are not in a field experience. Keep your instructor informed.

Activity 5: Learning about Liability

A. Write to the State Department of Education for information about O.S.H.A. (Occupational Safety and Health Act). Request any materials they provide for elementary school teachers. Examine the materials provided by your instructor to determine O.S.H.A.'s application to an elementary school classroom. What are your rights under this act?

B. List at least three common-sense guidelines for science teachers to guard against liability for negligence and to prevent injury to students.

Did you include careful planning in your list? If you did not, perhaps you should entirely re-read Part 5-2, "Safety in the Classroom."

C. What is the difference between tort liability and civil liability? What does *sovereign immunity* mean? Is sovereign immunity employed in your state? Find out.

D. Arrange for a seminar with your instructor to discuss O.S.H.A. as it relates to elementary school teaching.

Summary: Safety

The background information and the activities provided were designed to help you meet the behavioral objectives on page 266. Review these objectives to determine if you have reached competency in the area of safety in the science classroom. If you do *not* feel competent in this area, ask your instructor for individual help. If you feel that you have met the objectives, arrange for a final seminar with your instructor.

FINAL SEMINAR

Get together with a small group of your classmates and discuss the reading and activities done in this Part. Invite your instructor to join your discussion group.

NOTES

COMPETENCY EVALUATION MEASURE

Your instructor may choose to use a competency evaluation measure of some type to evaluate your competency in the area of classroom management. Check with your instructor for specific directions.

BIBLIOGRAPHY

Bozardt, D. Anita, and Righter, Roderick E. "The Supervisor and Teachers in Liability." In *Second Sourcebook for Science Supervisors,* edited by Mary B. Harbeck, pp. 171-76. Washington, D. C.: National Science Supervisors Association, 1976.

Brown, Billye W., and Brown, Walter R. *Science Teaching and The Law.* Washington, D. C.: National Science Teachers Association, 1969.

SUGGESTED READING

Brown, Billye W., and Brown, Walter R. *Science Teaching and the Law.* Washington, D.C.: National Science Teachers Association, 1969.

Irving, James. R. "How to Provide for Safety in the Science Laboratory." Washington, D.C.: National Science Teachers Association, 1968.

"Occupational Safety and Health Act: A Responsibility for Science Teachers." *The Science Teacher* 41, no. 7 (October 1974): 35.

A local school system's safety guide.

Your state's guidelines for O.S.H.A.

UNIT 6
Involvement Activities

Two Parts are included in this Unit: "Metric Measurement" and "Project Ideas." They were especially designed for preservice and inservice teachers who have very little background in these specific areas. Rather than exploring the subjects in depth, they aim to provide an awareness of each area. Moreover, they are practical in orientation rather than theoretical. Here are a few suggestions of ways to use these Parts:

1. They may be used as individual projects done independently outside of class.
2. They may be used as a basis for constructing activities that could be used with elementary children. These activities could actually be implemented and the results reported back to the class.
3. They may be used as part of the required material for the class.
4. You may do them just for fun, on your own, while taking the class or save them for later when you have more time.

There are many other ways that these two Parts can be used. You, your classmates, and your instructor can best decide how this material can help you become a competent elementary science teacher.

PART 6-1

Metric Measurement

FLOWCHART: Metric Measurement

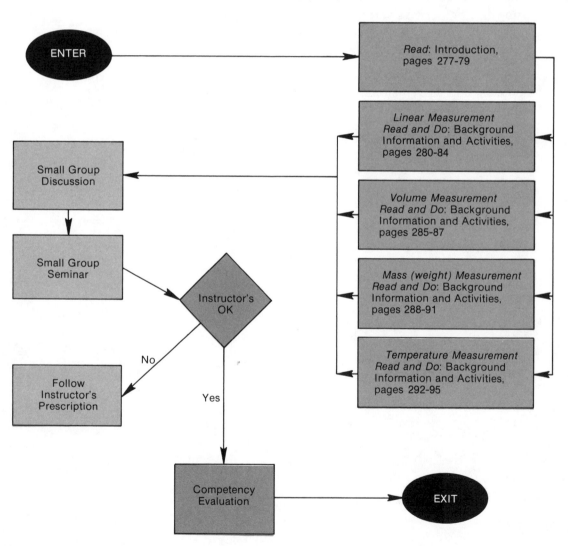

Introduction

Skill in measuring is essential to any effective science program. Even in the early grades, science activities involve students in collecting data because measurement is the tool that allows learners to describe and explain natural phenomena. It is used to make comparisons among objects and to determine how much, or how little, change has taken place over a specified time period. Furthermore, it allows students to make predictions and establish controls. Measurement provides students with a referent or standard which can be used to compare an unknown.

The process of measuring is a learned skill. Therefore, instruction must be planned that involves learners in this process. These planned experiences must take into account the students' physical, cognitive, social, and emotional development. Generally, the measuring process proceeds from very concrete comparisons to more abstract ones.

Part 6-1 is designed to introduce you to the metric system, a system of measurement that has evolved over the past 180 years. The Systeme International or SI, adopted in 1960, is the modernized version. It provides the basis for the units needed for everyday usage. The SI units of kilogram, metre, and litre are gradually replacing the familiar measures which were based on the English pound and pint.[1]

On December 23, 1975, President Ford signed the Metric Conversion Bill which calls for a gradual conversion to the metric system by 1985. Subsequently, preparing learners to deal effectively with a metric world has become a major responsibility for our educational system. The individual classroom teacher will have to assume the primary responsibility for implementing a practical sequence of activities for learning this system of measuring.

In this Part, you will be introduced to the overall process of measuring as well as to the specific units of the metric system. The background information will include a description of the relationship of the various metric units and will provide a rationale for learning this system. A teaching philosophy that is centered around "thinking metric" will be stressed.

The aim here is to provide preservice and inservice teachers with appropriate background information and practical activities so that they can begin to develop a common use competency in the area of metric measurement. This Part is divided into four sections: linear measurement, volume, mass (weight), and temperature. These four areas of measurement are most common in elementary school science. A fifth common area of measurement is time. It is not dealt with here because the units of time (seconds) will not be altered by the conversion to the metric system.

In each section, you will be given specific background information about that particular area of measurement. Terms that identify the appropriate metric units will be included, appropriate measuring devices will be examined, and opportunities for you to work concretely with the various metric units will be provided. Unlike previous Parts, the activities fall within the informational material instead of being in a separate section. This meshing of information with practical experiences should help you begin to *think metric*. It is hoped that this Part will provide you with a good foundation upon which you can continue to build in the area of metric measurement.

---

[1]The British spellings metre and litre have been adopted by many groups to encourage worldwide agreement in the spelling of metric terms.

GOALS

After completing this Part, you will demonstrate competency in the following:
A. The ability to estimate using common metric units
B. The ability to use appropriate measuring devices to quantify data in metric units
C. The ability to use the metric system of measurement in describing, explaining, and comparing natural phenomena

BEHAVIORAL OBJECTIVES

In completing this Part, you will do the following:
A. Identify common usage metric units
B. Estimate in appropriate metric units: the linear dimensions of a given object, the volume of a given liquid, the mass (weight) of a given object, and the temperature of a given substance
C. Use linear measuring devices which are marked in metric units in obtaining data about natural phenomena
D. Measure the volume of various amounts of liquids and other containers by using a container marked in metric units
E. Use a balance and standardized weights to measure in metric units the mass (weight) of various objects
F. Measure the temperature of various given substances using a thermometer with the Celsius scale
G. Describe, explain, and compare given substances and objects using metric units of measurement

Before you become involved with the four specific areas of metric measurement, you should receive some background information on the process of measuring. The simplest form of measurement involves concrete perceptual comparisons of some characteristic of an object to a known referent. Learners must have many concrete experiences in making perceptual observations with various objects. In the area of measurement, we are most interested in guiding beginning learners to observe characteristics such as length, volume, mass (weight), and temperature. Nonstandard referents are commonly used in the beginning stage. "As heavy as my pencil," "longer than my arm," and "colder than tap water" are some examples. As the learner matures and develops, measurement concepts begin to be more abstract. Standard referents with numerical representations are used to describe and compare observations: three metres long, 23° Celsius, between five and six grams.

It is important that teachers and learners alike understand the process of measurement. Simply put, a basic referent must be identified. For beginning learners and those who have not had appropriate experiences, the use of concrete, nonstandard referents are suggested in order to help the learner understand the process. A method must be devised for counting how many of the referents or what portion of one equals the characteristic of the object being measured. The unknown is then described by comparing it with the referent. The beginning learner uses terms such as *more, less,* or *the same as.* But as he gains experience and develops cognitively, he can make more precise measurements.

Many times, learners become skilled in measuring but are unable to make decisions concerning what to measure, which measurements to make, and how to best use the measurements once they have them. Teachers must help students establish a purpose for measuring. One way is by providing many science experiences in which measuring is useful in helping describe and explain natural phenomena.

Linear Measurement

Background Information and Activities

A. Each area of measurement in the metric system has one basic unit associated with it. The *metre* (m) is the basic unit for linear measurement. Larger and smaller quantities of the basic unit are indicated by using prefixes. A metre can be subdivided into smaller units of 1000 *millimetres* (mm), 100 *centimetres* (cm), and 10 *decimetres* (dm). The prefix *milli* means one-thousandth, *centi* means one-hundredth, and *deci* means one-tenth.

1. Obtain a metrestick and examine the metric scale marked on it. You should notice that the metre is divided into 1000 small units. Each of these units is called a millimetre (mm) and represents 1/1000 of a metre. After every group of 10 millimetres, there should be a numeral. This unit of length (10 millimetres) is called a centimetre (cm). There are 100 centimetres in a metre. These 100 centimetres can be divided into groups of 10. Each group of 10 centimetres is called a decimetre (dm). There are 10 decimetres in a metre.

B. It should be evident to you now that the metric scale is based on a decimal system. This logically conceived system has been designed so that all measures are related through a factor of ten. This means that conversion from any one unit to any other unit within the same area—millimetres to centimetres, for example—can be made by dividing or multiplying by ten or powers of ten. This is much simpler than our present system which employs a multiple of conversion factors. For example, in the English system of measurement, to convert feet to inches the factor is twelve. To convert feet to yards the factor is three. With each conversion, a different factor is necessary. In the metric system, the conversion factor is always ten or powers of ten.

One important thing you must remember in the metric system is the order in size and the relationship of the units. The following shows the order in size of the units mentioned so far and their relationship.

1 metre = 10 decimetres = 100 centimetres = 1000 millimetres

C. In order to convert from centimetres to millimetres, you would multiply by ten because you are going from a larger unit to the next smaller unit. You would divide by ten to convert centimetres to decimetres because you are going from a smaller unit to the next larger unit.

1. The following questions will give you a chance to use the information just presented:
    a. How many centimetres are in 60 millimetres?
    b. How many millimetres are in 15 decimetres?
    c. How many decimetres are in 60 centimetres?
    d. How many decimetres are in 640 millimetres?
    e. How many millimetres are in 4.5 decimetres?
    f. How many centimetres are in 1000 millimetres?
2. Use a metrestick to check out your answers. You might want to compare them with your classmates' answers. If there are any conflicts, try to resolve them. Ask your instructor for help if you have trouble.

D. There are three other common linear metric units with which you will want to be familiar. These units are all larger than the basic unit of the metre. The *decametre* (dkm) is equal to 10 metres; the *hectometre* (hm) is equal to 100 metres; and the *kilometre* (km) is equal to 1000 metres. With the metric system, the distances between cities will be measured in kilometres.

E. The following activities are designed to give you some concrete experiences in using linear metric units.

1. Look around you and find at least three objects that you estimate to be about one metre in length, width, or height. *Do not measure these.* Record your answers.
    Objects estimated to be approximately one metre:
    a.
    b.
    c.
2. Obtain a metrestick and check your estimates by comparing them with the metrestick. Record the results as being more than a metre, less than a metre, or about the same as a metre.
    Results
    a.

    b.

    c.

3. See if you can find three objects that you estimate to be about equal to two metres. Record these below.
    Objects estimated to be approximately two metres:
    a.

    b.

    c.

4. Use the metrestick and check your answers. Record the results as being more than two metres, less than two metres, or about the same as two metres.

Results

a.

b.

c.

5. Try to estimate the length and width of your classroom in metres. Then use a metric tape to check your answer. Record your results.

Estimate   *Actual Measurement*

6. Try some other estimating in metres, then measure to see how accurate you are.

Estimates   *Actual Measurements*

F. Activities involving estimating and measuring objects of one metre are best to use in beginning measurement experiences. Once the metre length has been conceptualized, other activities that focus on estimating and measuring larger and smaller units of the metre are needed. By associating the various metric lengths with the length of common things, conceptualizing the units becomes easier.

G. The following activities involve you in estimating quantities smaller than a metre:

1. List three things that you estimate to be approximately equal to one millimetre.

a.

b.

c.

2. List three things that you estimate to be approximately equal to one centimetre

a.

b.

c.

3. List three things that you estimate to be approximately equal to one decimetre.

   a.

   b.

   c.

4. Use a metrestick as a guide and mark off a strip of stiff paper into ten centimetres. Divide each centimetre into ten equal units to represent millimetres. Use this ten centimetre strip to check the estimates you made. Indicate next to your answers above whether your estimate was more, less, or approximately equal to the specific unit indicated.

5. Compare your results with others. See if you can find other things that are approximately equal to a millimetre, centimetre, and decimetre. What different parts of your body are approximately equal to any of the units?

H. The larger quantities of distance measurement—decametre, hectometre, and kilometre—do not lend themselves to concrete, inclass activities; so no planned activities involving these larger units are included here. However, some suggestions are offered that could be used to help you conceptualize these larger units. Activities outside the classroom involving cars, bicycles, or track and field events can be used to acquire a concrete idea of these longer distances. Making a scale drawing of a neighborhood, university campus, or elementary school playground using metric units is also helpful.

Many different kinds of materials can be used in constructing metric tape measures—ropes, adding machine tape, plastic webbing, drapery tape. These materials can be marked off in metric units (generally one metre lengths with decametres and hectometres indicated) using a commercial tape as a guide.

You can probably think of many other ideas for teaching the metric system. Share these with others interested in becoming more familiar with the metric system. You might want to start a file of ideas to use when you work with elementary children.

Avoid converting from the familiar English units—mile, rod, acre—to metric units with elementary children. It is much better to give them many concrete experiences so that they can think in metric terms.

I. To get more practice in using linear metric units, measure the following items using the metrestick for large measurements and the ten centimetre strip to make smaller ones. Record your answers.

1. The length of your shoe in centimetres.

2. The height of an average door in metres.

3. The width of a desk in decimetres.

4. The thickness of a nickle in millimetres.

5. The length of your outstretched arm in centimetres.

6. The length and width of your bed.

7. The length and width of your room.

8. Make several other measurements and record these below.

    a.

    b.

    c.

    d.

    e.

J. You should be well on your way to acquiring a concrete understanding of the common metric units of linear measurement. Jot down any comments or notes you wish to make. If you feel you need more feedback or help, ask your instructor to discuss this portion with you.

COMMENTS

Volume

Background Information and Activities

A. The metric unit for describing the volume of liquids and gases is the *litre* (l). The litre is composed of 1000 equal subunits. Each subunit is called a *millilitre* (ml). The millilitre is the referent for volume and has a logical relationship with linear measurement units. One millilitre can be described in linear terms as one cubic centimetre (1 cm^3). This means that volume—how much space something occupies—is directly related to linear measurement. One millilitre is equal to 1 cm x l cm x l cm, or l cm^3.

These students are learning how volume relates to linear measurement.

Since there are 100 millilitres in 1 litre, and each millilitre is equal to 1 cubic centimetre, a litre would be equal to 1000 cubic centimetres or 1 cubic decimetre (1 dm^3).

B. Obtain several graduated cylinders and examine them carefully. Notice the scale markings on each cylinder. These markings indicate the measurements intervals. Some cylinders have a double scale marking with the zero point at the top on one side and at the bottom on the other. Generally, 50 ml and 100 ml cylinders are graduated in 1 ml intervals.

1. Determine the volume of liquid each interval on the scale measures for three different sized cylinders that you are examining. Record your findings.

Maximum volume measured by selected cylinder	*Volume of liquid each interval on scale measures*
a.	
b.	
c.	

C. Obtain a litre box and examine it carefully. The inside dimensions of the litre box are 1 dm x 1dm x l dm, or 1 dm³. You can verify this by using a metrestick or a 10 cm strip. Most litre boxes are graduated in 100 ml intervals. The litre box can help you visualize concretely the relationship between linear measurement units and volume measurement units.

1. Select one of the graduated cylinders and a litre box. Make several practice measurements by pouring different amounts of water into each. Be sure the top surface of the water is at eye level when you make a reading. If you are using a glass cylinder, you will notice that the top surface of the water curves downward from the sides of the container. Water particles tend to cling to the glass sides of the cylinder which causes the surface to be curved. In order to obtain an accurate reading, this curve, called the *meniscus,* is kept at eye level and the lowest part of the downward curve is used in making the measurement. If you are using a plastic cylinder, the surface should be level and not curved downward. Nonbreakable plastic cylinders and litre boxes are recommended for elementary classrooms. The measurement intervals should be clearly marked and easy to read.

2. Select several of the containers provided and use the graduated cylinders to determine the amount of liquid each will hold. *Suggestion:* Fill the graduated cylinder with water and note the amount. Pour the water from the graduated cylinder into the selected container, instead of filling the container and pouring the water into the graduated cylinder. When the water level is flush with the top of the container, the container is considered full. Record your findings.

Container selected	*Amount of liquid held when full*
a.	
b.	
c.	
d.	

3. Now that you have had some experience with measuring volume using metric units, try to estimate, in metric units, the amount of liquid some familiar containers will hold when filled. Some suggestions are listed to

get you started. Space is provided for listing others of your own choosing. Record your estimates and the actual volume measured.

Description of container   *Estimate of volume*   *Actual volume*

a. a juice glass

b. a coffee mug

c. a medium size pitcher

d.

e.

f.

g.

4. You may want to calibrate some familiar containers in metric measurement intervals. This can be done by using the referent of 1 ml. Pour 1 ml of water into the selected container and mark the water level. Continue this procedure until you reach the desired volume. Larger containers can be marked in 100 ml intervals. These "home-made" graduated containers can be used effectively with elementary children to involve them with metric measurements. Older children can construct their own graduated containers to use in measuring volume.

5. The volume of solids can be measured by displacement. The cubic centimetre (cm^3) is usually used instead of the millilitre (ml) in describing solid volume measurements. Practice making volume measurements using the solid objects provided. Pour a measured amount of water into a graduated container. Place the solid object into the container of water and measure the water level. The amount of water displaced—the second reading minus the first reading—will equal the volume of the solid. Record your findings.

Solid object used	*Volume of water before adding object*	*Volume of water after adding object*	*Volume of solid object*
a.			
b.			
c.			

D. With a group of your classmates, discuss the activities done in this section. Compare your findings and talk about any difficulties you had in understanding the information presented or in doing the activities suggested. Jot down comments, questions, or notes concerning volume measurement. If you feel you need more guidance or feedback, ask your instructor to join your discussion group.

COMMENTS

Mass (Weight)

Background Information and Activities

A. In the elementary grades, usually no distinction is made between the terms *mass* and *weight*. In this section, you will see the term *mass* followed by the term *weight* in parentheses. The mass of an object is the amount of matter or material it contains. The weight of an object is the amount of force being exerted on an object. The force being exerted on an object is determined by the pull of gravity. The following example will illustrate how the two measurements can differ. On the moon, the force of gravity is much weaker, about one-sixth the force on the earth. Because the force of gravity is less on the moon, the weight of an object would be less if measured on the moon than if measured on the earth. Mass is not affected by the force of gravity. Therefore, the mass of an object would always remain constant whether measured on the moon or on the earth.

The term *newton* (N) is used as the standard metric unit to indicate the weight of an object. A spring balance with either a rectangular or dial-shaped face, on which the graduated scale is located, is used for measuring the weight of objects. When an object is attached to the spring scale, the spring will be stretched according to the amount of force the object exerts. In the metric system, one unit of force is equal to the distance of one centimetre.

Small groups of students are learning about mass (weight).

The *kilogram* (kg) is the common unit of measurement for mass. A kilogram is composed of 1000 equal units. Each of these units is called a *gram* (g). Even though grams are used to indicate the mass of an object, they are also commonly used to indicate weight. This is acceptable because there is a direct relationship between the mass and weight of any given object. As long as the gravitational pull on an object is constant, the weight of the object is directly proportional to its mass. This means that an object with more mass will have more weight and one with less mass will have less weight. Therefore, for general classroom use, the terms *mass* and *weight* can be used to indicate either.

An equal arm balance and standard masses or weights are used to measure mass (weight) of objects. All operate on the principle of the lever or teeterboard. An object is placed in a pan which is attached to one side of the basic beam. At an equal distance from the center balancing point on the other side is another pan attached to the basic beam. When a second object of equal mass is placed into the pan on this side, the basic beam should return to a horizontal or balanced position. If the two objects do not have the same mass, the beam will slant downward toward the object with the heavier mass. A spring balance with a scale calibrated in grams rather than newtons can also be used for measuring mass (weight).

1. Obtain several equal arm balances and examine them. Practice using one of the balances by selecting an object and placing it on one side of the balance. Try to find another object that has the same mass as the first. Place the second object on the other side and see if the balance returns to the horizontal position. Continue trying various objects until you find one with the same mass as the first. Next, try to balance the first object with two or three objects.

B. Now that you have had experience in using the equal arm balance, let's examine the standard masses or weights that are used to measure mass (weight). As you learned earlier, the gram is the unit for measuring mass (weight). This unit was determined by weighing one millimetre of cold water. Therefore, one gram is equal to the mass of one millimetre of cold water. Remember that 1 ml is equal to 1 cm³, or 1 cm x l cm x l cm. This logically conceived relationship among mass (weight), volume, and linear measurement allows for easy movement from one measurement to another. The following illustrates the relationship among volume, linear, and mass (weight) metric measurement units.

$$1 \text{ cm}^3 = 1 \text{ ml} = 1 \text{ g}$$

1. Obtain a set of standard masses or weights from your instructor. There are several different kinds. The plastic stacking discs and cubes are appropriate for the early elementary grades. They are relatively inexpensive and accurate enough for early experiences in measuring mass (weight). For the upper elementary grades, a brass gram mass set is more appropriate. These sets are more expensive but are much more accurate. Place one of the mass pieces in your hand and feel the weight. Try to find an object in the room that feels to be about the same weight. Put the selected object on one side of a balance scale and the standard mass on the other. Are they the same weight? Try this procedure with several other standard mass pieces and objects.

2. Find the mass (weight) of the following items. Also add some items of your own choosing.

Items	Measured mass
a. Your pencil	
b. This textbook	
c. An empty soft drink can	
d. A piece of chalk	
e. A nickel	
f.	
g.	
h.	

3. If there are some spring balances available, you might want to use them for practice in measuring the mass (weight) of several selected objects. Just attach the object to be measured to the spring scale. The measurement units are located on the face, and a marker indicates the mass (weight) of the object. Record your findings.

Items	Measured mass
a.	
b.	
c.	
d.	

C. If a metric bathroom scale is available, weigh yourself. You will find that your weight is measured in *kilograms*. Compare your weight with that of several other people. After you have some experience with this, try to estimate the weights in kilograms of several people of different sizes; then compare your estimates to their actual weights.

If a metric scale is *not* available, you can convert one that measures in pounds by recalibrating the scale. To get to the scale, you must remove the plastic covering and the top. The top is usually held in place by very strong springs, which must be pulled up to release their hold on the top. Once the top is off, remove the dial plate and glue a clean piece of paper over the old scale. Mark off the new scale in 5 kilogram intervals. Each 2.2 pounds equals 1 kilogram; therefore, 11 pounds would be equal to 5 kilograms; 22 pounds is equal to 10 kilograms; 33 pounds is equal to 15 kilograms, etc. When the new scale is finished, replace the top, making sure you get each spring through the top.

D. Discuss these activities with a small group of your classmates. Compare your findings and talk about difficulties you had in understanding the information presented or in doing the activities suggested. Jot down comments, questions, or notes concerning mass (weight) measurement. If you feel you need more guidance or feedback, ask your instructor to join your discussion group.

COMMENTS

Temperature

Background Information and Activities

A. The unit for measuring temperature is a *degree*. The term *degree* is used in both the familiar Fahrenheit (F) scale and the metric scale, Celsius (C), but the unit of measurement a degree represents on the Celsius scale is *not* the same as a unit of measurement on the Fahrenheit scale. The Celsius scale, which was called the centigrade scale for a number of years, has 100 degree units. The two extremes, 100° and 0°, represent the boiling point and freezing point of water under standard pressure conditions.

These children are using a thermometer to learn about temperature.

A thermometer is used to measure the temperature of an object or system of objects. Temperature can be thought of as the degree of hotness or coldness of things. Young children can be helped to develop the concept of temperature by being involved in concrete experiences in which they are able to "feel" the hotness or coldness. For example, several containers of water can be used—one warm, one cold, one very warm. The learners can use the sense of touch to feel the temperature of the water, and then a thermometer can be used to quantify the observations and to help the children understand how the thermometer operates. For instance, a child puts his hand into a container of cold water. Then the thermometer is put in and a reading made. This reading can be associated with the feeling of coldness the child experienced when his hand was in the water. The child can also observe that the red liquid goes up when the thermometer is placed in warm water and goes down when placed in cold water. Many concrete experiences such as these are necessary for children to develop the concept of temperature.

1. Obtain a thermometer with a Celsius scale. Examine it carefully. Notice the degree markings on the side of the scale. Generally, student thermometers with a C scale are marked in one degree intervals, which are easier for elementary children to read than those on the F scale, which are marked in two degree units.

2. One of the best ways to begin to think about temperature in metric terms is to establish a reference point. Room temperature serves this purpose well. Place your thermometer in a spot that you feel will give the most accurate reading of the temperature in the classroom. If possible, suspend the thermometer to minimize its contact with objects. This contact can some-times affect the reading. Wait about three minutes before taking a reading. If you feel that thermal equilibrium—the point at which the liquid column becomes stationary—has not been reached, leave it for two additional minutes and make a second reading. Record your finding.

3. Now that you have measured the room temperature in degrees Celsius, make some observations that you can associate with that reading. These observations will help you begin to Think Metric. Use the following questions to help you in making some observations:
   a. How do you feel? Warm? Cool? Comfortable?
   b. What kind of clothing is most appropriate for staying comfortable in the temperature recorded?
   c. Are there other observations you can make to associate with the recorded temperature?

B. Room temperature can provide a reference point for thinking about temperature in metric terms. Generally, comfortable room temperature is approximately 22°-23°C. Consequently, temperatures higher than that will be warmer, and temperatures lower than that will be colder. It is a good idea to establish the degrees Celsius that you consider cool, cold, and hot and to make observations associated with each.

1. Use the thermometer to measure the temperature of the following in degrees Celsius.

Items	*Temperature*
a. Tap water	°C
b. Warm water	°C
c. Water from water fountain	°C
d. Outside air	°C
e. Mixture of ice and water	°C

2. Make some other measurements of your own choosing using the Celsius scale.

Items	*Temperature*
a.	
b.	
c.	
d.	

C. All of us will need many concrete experiences in which we can associate the various degrees Celsius with our own observations. Activities for children should involve them in making observations about themselves in relation to the temperature such as the following:

1. How do I feel in this temperature, cold? hot? cool? warm? comfortable?
2. What clothes do I feel most comfortable wearing in this temperature?
3. Can I see my breath in this temperature?
4. Should I wear gloves in this temperature?
5. Do I perspire easily in this temperature?

D. You might want to make a metric profile for yourself. Below are some suggestions.

height	_____ cm
weight	_____ kg
body temperature	_____ °C
neck	_____ cm
shoulders	_____ cm
arms	_____ cm
wrist	_____ cm
chest	_____ cm
waist	_____ cm
hips	_____ cm
legs	_____ cm
calf	_____ cm
feet	_____ cm

E. Discuss these activities with a group of your classmates. Compare your findings and talk about difficulties you had in understanding the information presented or in doing the activities suggested. Jot down comments, questions, or notes concerning mass (weight) measurement. If you feel you need more guidance or feedback, ask your instructor to join your discussion group. Remember, Think Metric!

COMMENTS

Summary: Metric Measurement

The metric system of measurement is really very simple to learn and use. It is less complex than the English system and much more logical. The best way to become familiar with the system is to think in metric terms. The terms and the type of measurement they represent must be committed to memory—metres for linear measurement, litres for volume, gram for mass (weight), and degrees Celsius for temperatures. The prefixes of *milli* for one-thousandth, *centi* for one-hundredth, *deci* for one-tenth, and *kilo* for one-thousand are used to indicate measurements that are more or less than the basic unit. Remember that all the metric units are related through a factor of ten. Linear, volume, and mass (weight) units are all related in the following way:

$$1 \text{ cm}^3 = 1 \text{ ml} = 1 \text{ g}$$

The best way to learn the metric system of measurement is to *think metric. Don't convert* the familiar English units to metric units or vice versa. Try to visualize metric units in terms of familiar objects, not as so many inches, pints, or pounds. Make measurement in metric units as often as possible. Estimating before you measure is also helpful. For everyday use, the units that you worked with in this Part will serve you well. There are several publications available that give a more complete and in-depth description of the metric system. These are listed in the bibliography. Furthermore, many states have a metric resource guide available for teachers. Many school systems are beginning to develop their own metric resource materials. You will want to begin *now* to think metric so that you will feel comfortable in helping elementary students to learn and use the metric system of measurement.

FINAL SEMINAR:

Review all portions of this Part. Look over the questions and comments that were recorded after the various activities. Get together with others who have completed the Part and discuss your reactions. If you have specific questions, share them with the group; together you might be able to answer them.

Arrange a final group seminar with your instructor. Be prepared to share your individual and group reactions to the information presented and activities completed. Record your notes from the seminar.

NOTES

COMPETENCY EVALUATION

The activities that you completed in this Part are considered appropriate practice. Your instructor may choose to use a competency evaluation measure to evaluate your competency in the area of metric measurement. See your instructor for specific directions.

BIBLIOGRAPHY

Carin, Arthur, and Sund, Robert. *Teaching Science through Discovery.* 3d ed. Columbus, Ohio: Charles E. Merrill Publishing Co., 1975. Chapter 14.

Rowe, Mary Budd. *Teacher Resource Guide for Metric Education.* Lansing, Mich.: Michigan Department of Education, 1976.

————— . *Teaching Science as Continuous Inquiry.* New York: McGraw-Hill, 1973. Chapter 2.

SUGGESTED READING

The following publications are provided by the American National Metric Council, the American National Standards Institute, and the National Bureau of Standards.

ISO Recommendation R 1000, *Rules for the Use of Units of the International System of Units,* 21 pp. American National Standards Institute, 1430 Broadway, New York, N.Y. 10018

ANMC Metric Editorial Guide—Interim Guide to Accepted Metric Practice, 11 pp. American National Metric Council, 1625 Massachusetts Avenue, N. W., Washington, D. C. 20036

NBS Guidelines for Use of the Metric System, November, 1974. Metric Information Office, National Bureau of Standards, Washington, D. C. 20234

PART 6-2

Project Ideas

FLOWCHART: Project Ideas

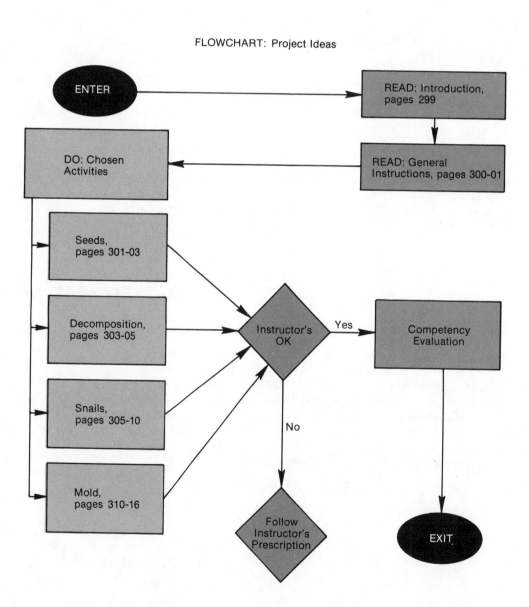

Introduction

Young children are fascinated by living things, whether plants or animals. They like to see them grow and change form as they mature. Almost all elementary teachers have something alive in their classrooms (students are excluded). Sometimes the room may resemble a zoo, at other times an arboretum. Sometimes there may be only one plant, but almost always, there is something living in the classroom. In this way, teachers utilize whatever is available to teach students about living things. The purpose of this Part is to acquaint you with some projects that you might use with your students, as well as to give you firsthand experiences in observing and reporting.

Involvement projects are fun activities designed to give you an opportunity to do some life science activities on your own that you probably would not ordinarily think of doing. Most of the activities that you have done or will do in the various Parts of this book are centered around the physical sciences for two obvious reasons: physical science activities can be done more quickly than life science activities, and the equipment is easier to maintain and store. It does take time for something to grow or to collect observational data. That is also the reason why any, or all, of the activities presented here should be started early in the semester. You will need at least three or four weeks to really get much out of them.

Four activities are presented here. The format for each will be slightly different, ranging from a nonstructured set of directions to a very complete, record-keeping set of directions. This will allow you to learn several ways of presenting activities to your students.

You may want to do one, two, three, or all four of these activities. Each one is completely different from the others so that learning will not overlap if you wish to do them all. Your instructor may choose to require one or more of these activities, and if this is the case, follow her requirement. If you are working on your own, do as many as you like.

Read the general instructions before attempting the activities. Doing so will save you some embarrassment later, especially in reporting your work.

This Part will get you to look closely at some life science phenomena that you may have seen only superficially. If you really want to learn about something, get involved. Here is your chance.

GOAL

After completing this Part, you will demonstrate competency in the ability to set up and carry out an experiment on your own, without instructor help.

BEHAVIORAL OBJECTIVES

In completing this Part, you will do the following:
A. Devise an experiment and carry it out
B. Make valid observations about living organisms
C. Identify and explain one or more ways to present an activity to children based on the degree of structure given and information requested

General Instructions

You will find that if you really get involved, you will enjoy these activites and will learn a lot of interesting things that you didn't know before. You probably have read about most of the activities suggested or have seen them casually but have never really taken a close look at them. Now is your chance to actually watch a seed break through the ground or see snail eggs as they get ready to hatch, or some other equally fascinating phenomena. The key word in this Part is *involvement*. You can read for background information or to help you understand your observations, but you cannot read for the answers—you must experience them. You will get out of this Part only what you put into it. With this in mind, here are some general instructions for all the activities. Specific instructions relating to each activity will be given with the activity.

1. Read all instructions before you begin. Ignorance of the directions is no excuse for not completing an activity to the satisfaction of the instructor. Incomplete activities result in incomplete credit from your instructor. If, after reading the instructions, you still have questions, ask for help.

2. Select the activity or activities that you are going to do. Don't do activities that you have already done in another class because you already know what will happen. Try something new. Be adventurous. Don't turn up your nose at mold simply because you have always thrown moldy food away when you found it in your home. Take time to really look at some mold. Have you ever really seen red mold? or yellow? It is quite pretty, really. Look at all four activities before you decide which one(s) you want to do.

3. Look at the format of each of the activities. Each one is different. The reason for using the specific format is explained in the introduction to the activity. By reading the materials given for each activity, you will find out several different ways that you can structure an activity for your students.

4. Do the activity. You are to work alone, not with partners or a group. However, you can talk to others, compare your project with theirs, and swap information. You will have to obtain your own materials. Your instructor may or may not provide some of them, but involvement includes gathering materials.

5. Allow yourself plenty of time to complete the activity. You cannot do any of these in a few days. Think in terms of weeks, usually about four or five, with a very minimum of three. Mold may be the exception, depending on how fast it grows.

6. Success or failure? What happens if nothing happens? This is where the actual learning takes place in these activities. If everything goes well, you watch and wonder, then report what happened. If not, then you have a better learning opportunity than those who have immediate success because you must find out why and then try the activity until you are successful. If your seed, for example, doesn't grow, find out why and plant another, and yet another if need be. If you were to use any of these activities, or variations of them, in your classroom, you would have to know what to do if nothing happened. You must have success in these activities, and sometimes the greatest success arises from initial failure. One exception might be with your snails. If they don't lay eggs, we don't know how to make them. But, you can check around for information about how, when, and under what conditions they are supposed to reproduce and report that.

7. You are to report your activities to your instructor in two forms: written and oral. The written report will follow the format established by the particular activity but is not limited to that format. In other words, you are expected to do more than just fill in the blanks. The blanks, where there are some, only report the factual observations of the process being carried out. *Use your imagination to supplement your report.* Make it interesting for you and your instructor. Then, present information about the activity as a teaching tool. How could you use it in a classroom situation? Did you enjoy it? Would children enjoy it? What were your reactions? Did you change your outlook on the subject of the activity? Think about what you have done, and the implications for teaching. Then write it down before you forget it.

The second portion of the report requires that you present your written report and discuss it. You may be asked to bring your plants, mold, etc., along with your report. Check with your instructor about this requirement. Now is your chance to show off your project and to ask some of those questions that came up but couldn't be answered.

If you have read these general instructions thoroughly, you are ready to start the activities. Remember, they are to be done out of class, on your own.

Background Information and Activities

Seeds

This activity uses seeds as a method of involving students in observing how a plant grows. Seeds are planted, watered, and observed. Various experiments can be carried out on the plants to verify some of the things that you have read or heard about, such as how light affects plants or how much water plants need. This will give you some experience in formulating a question, devising an experiment, and then carrying it out to answer your question. Remember to control all of your variables when you do an experiment. You cannot start two plants, than put one in the light and water it regularly and put the other in the dark without water and expect to get valid results. Do you know why not? If you don't, you should go back to some of the previous Parts, for instance the SAPA II activity on variables on pages 118-22 and find out.

You will be given only a minimum of structure in this activity, and it will be up to you to set up your experiment, explain it, carry it out, and report it. No format for reporting your data is given. You will have to devise your own. You might want to present your data as you would to a group of first grade students or to college biology students, or somewhere in between. The choice is yours. Two important items to remember:

1. You must devise a written report of your experiments and observations. Your instructor might also want to see your actual plants, so check before you get rid of them.

*An excellent involvement activity consists of giving students the
opportunity to observe plant growth.*

2. Use your imagination when reporting. Don't use the old day-by-day diary
   approach. Be creative! Draw pictures, make graphs, write stories, and
   anything else you can think of.

EQUIPMENT

Seeds—a bean or pea works well, but others, such as flower or vegetable
seeds, also work

Soil—your choice, but commercial potting soil usually works best

Container—Styrofoam cups work well, but clear plastic cups allow you to see
the root structure

Hand lens (magnifying glass)—optional, but makes the observations more
interesting

Water pitcher—you have to have something to water the seeds with

1. Punch a small drain hole or two in the bottom of your cup(s). This is very important to keep the seeds from rotting.
2. Fill the cup(s) with soil. Pack it lightly.
3. Plant two or three seeds in each cup. Don't plant too deep.
4. Water. A small amount might be better than a deluge. (This might be a variable worth investigating.)
5. Start your observations. Don't forget to keep records.

VARIATIONS

You should do some experimenting with your plants. You may vary light, water, or temperature, but only one at a time. Remember variables? You may want to start your seeds between damp paper towels to see germination take place. Extra seeds may be started this way, then cut open for examination. Do be sure to record your variations and the observations you make.

OBSERVING AND REPORTING

Prepare your report on separate sheets of paper, and remember to make it interesting. Be sure to include a section that gives your feelings and observations about this project. Did you like it or not? What would you do next time? How could you use this activity with children? Was it worthwhile? *Make an appointment with your instructor* to discuss this activity.

COMMENTS

Decomposition

Decomposition is one of the natural phenomena that people depend on very much but know very little about. "Ashes to ashes and dust to dust" refers to almost everything in nature, for after an object or organism has served its useful purpose, it eventually returns to the earth from whence it came. We accept this as a fact, but since most decomposition takes place slowly and/or out of sight, we are usually not aware of its taking place. Sometimes there is an exception to this rule, and we are made aware of it by our noses. But, most decomposition takes place without our knowledge. In this activity, you will get the opportunity to experiment with decomposition and to become more aware of what materials will or will not decompose and how fast decomposition takes place. Don't be squeamish and overlook this activity; it is really quite interesting.

*One life
science
activity could
be the study
of decompo-
sition.*

You will be given slightly more structure in this activity than you were given in "Seeds." It will still be necessary, however, for you to enhance the activity and your presentation by thinking of variations that you might try.

EQUIPMENT

For this activity, you will need four specific items and six or more items of your own choice.

1. A container at least 15 cm x 30 cm x 10 cm, or 6″ x 12″ x 4″ if you must use inches. This container should be reasonably sturdy and relatively water-tight. A plastic shoe box is excellent, but a cardboard one can be used if you line it with aluminum foil.
2. Soil to fill the container almost full. This should be dirt from outside, if possible, rather than commercial potting soil. Try to avoid beach sand or hard clay if you can. You might want to try different types of soil in different containers to see what happens.
3. Small wooden stakes. Popsicle sticks or small tree limbs work well to hold the paper markers that you will use to help keep track of your objects.
4. Water to keep the soil moist. Don't flood the soil but keep it slightly damp. Try not to let it dry out completely.
5. Items of your own choice. You should choose a mixture of objects for this activity. Several possible choices are listed, and you should think of some of your own. Here are a few suggestions: a leaf from a tree; a sugar cube; a piece of bark, fruit, meat, iron, aluminum, or plastic.

PROCEDURE

This activity will take at least three to four weeks to complete, and it cannot be hurried; so don't get impatient.

1. Fill the container to a depth of about 7 cm with soil.
2. Bury each of the items that you have selected about 3 cm deep in the soil.
3. Mark each item with a stick. Use a piece of paper or masking tape on the stick to identify each item. This step is very important. You must know what you buried and exactly where you buried it. This will be your marker.
4. Moisten the soil until slightly damp. Don't saturate the soil. Check periodically and add small amounts of water as needed.
5. Put the container in a safe place, preferably out of direct sunlight, and leave it undisturbed, except to add moisture.
6. Read some materials about decomposition while you are waiting. See if you can find out what to expect when you start to do step 7.
7. At the end of three weeks or more—three being the very minimum—check your box. Take a marker out, then uncover the item that you buried without disturbing the others. Record what you find. Do the same with the others.
8. Make a presentation for the activity. You will have to decide the format that you will use. Be imaginative. Use pictures, verbal descriptions, or whatever you can think of to make your presentation as informative and interesting as possible.
9. *Make an appointment with your instructor* to present your report. Make it interesting for both of you.

COMMENTS

Snails

How many times have you seen a snail in an aquarium? Did you really look at it, or did you just see it? Have you ever wondered how a snail moves or eats? Probably not, because you never really had any interest in it. You know that snails are put into aquariums, but for what purpose? You watch the fish swim around, but does anyone ever watch the snail move around? You are going to, if you decide to do this activity.

Snails in an aquarium are interesting organisms for observation.

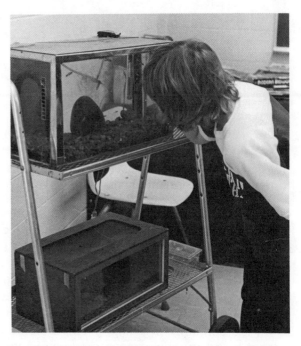

This activity focuses on snails and their reproduction. You will learn to identify the parts of a snail, the function of each part, and where and how to locate snail eggs. You will also have to do some reading for background information. A unit of this type could be used to help the children learn to observe, to use a hand lens, to keep accurate records, and to understand that animals' offspring are like the parent animal.

You will be given a more structured format in this activity than was found in the two previous activities. Specific questions with blanks for answers are part of the activity. However, you can formulate and try other experiments and answer questions of your own making as a supplement to the activity, and you should do this to enhance the activity. Add these variations to your report. Allow yourself at least four or five weeks for the completion of this activity. You may finish it much earlier if your snails lay eggs quickly or if you find some eggs that have already been deposited on a leaf or in the container.

EQUIPMENT

Snail(s)—These may be obtained from a tropical fish dealer, a friend with an aquarium, or maybe from your instructor. Try to get a large water snail. Very small snails are hard to observe.

Container—A mason jar with a top works very well. Punch holes in the top for ventilation. You do need a top because snails can crawl out. Almost any container will work if it is transparent.

Water—*Important!* Be sure to use aged water (water that has set out overnight in a container) to eliminate the chlorine in the water and to allow the water to reach room temperature. Small quantities of water can be added as needed, but it is still best to age the water first, if possible.

Hand lens—This is an absolute necessity in this activity for close-up viewing. You should have one in your possession at all times since you never know when you might come across something interesting to look at. Your instructor may have one that you can borrow, or you may have to furnish your own.

Food—Snails do eat, and you have to feed them. A small piece of lettuce is a good start. You might experiment with the snail's eating habits or look up that information in a reference book.

PROCEDURE

Use your hand lens to observe your snail, then fill in the first few blanks. This will help you get acquainted with the anatomy of the snail as well as some general characteristics. Read some information about snails. You will have to locate your own references. Then, based on some interesting fact that you read, develop and carry out an experiment with your snail. Meanwhile, look for snail eggs in your container. When you find some, mark them so that you can observe them, then record the information in the appropriate place on the chart. The following pages will constitute your report, but you can supplement it with additional sheets if necessary.

A. Getting Acquainted with a Snail

1. Draw a picture of your snail. Make it big enough to show the parts.

2. Label the parts of your snail. Find the feelers (or antennae), shell, foot, eyes, mouth. Can you think of any other parts that need labeling? If so, be sure to label them on your picture.
3. How do snails move around? Describe the motion.

What happens if the snail is disturbed?

4. Describe the habitat of your snail.

What is the ideal habitat for your snail?
(Answer this question before reading Part B of this activity.)

5. How do snails eat? Draw a picture of the mouth, then describe how eating takes place.

B. Reading about Snails

In this section, you will have to do some research. Go to the library and obtain some information about snails. You can use elementary or college level resources, but try to find out as much as you can about a snail. Here are some suggestions to guide your research:

1. Describe a snail verbally. How many different types are there?

Identify your snail scientifically and by common name.

2. Discuss the life cycle. How long does a snail live?

3. What do snails eat?

4. What is the ideal habitat for snails? What happens when conditions are not ideal?

5. How do snails reproduce?

C. Baby Snails

Most snails lay eggs, and if you can find them, you can watch them develop and hatch into baby snails. Look on the sides of your container or on floating material in the container. You will need a hand lens for this activity unless you have exceptional eyesight. Look for a small, clear mass, with a small black dot in each cell-like division. It will probably be a spot about one centimetre in diameter. This description fits some of the more common water snails and may not fit your snail. Refer to the research that you did in Part B of this activity for more specific information. Keep a record of your observations using the chart below.

Week	Mon	Tues	Wed	Thurs	Fri	Sat	Sun
1							
2							
3							
4							
5							

1. When did you first start observing your snail? Color the day in green.
2. When did you first see snail eggs? Color the day in red (a red-letter day). Describe what you found.

3. What changes did you observe in the eggs between the time that you found them and the time they hatched?
   a.
   b.
   c.
4. When did the eggs hatch into baby snails? Color the day in pink and blue. Then, describe the baby snails.

D. What Good Is This?

You have studied a snail, and you may or may not have seen baby snails hatch. So what? Remember, the purpose of this Part is to give you some help in becoming a better teacher, not to make a "snail expert" out of you.

E. Reporting the Activity

Make an appointment with your instructor to report what you have done in this activity. Bring in any supplementary material you have to go with this written work. Make the presentation interesting and informative.

COMMENTS

Mold

How many times have you found mold on something in your refrigerator or on a loaf of bread? What did you do? Did you look at it closely, or did you immediately throw it out? Generally, mold is not considered to be a very pretty sight, probably because it has spoiled something edible. But if you really take time to look at it, it can be very pretty. It is also extremely important in our ecosystem.

In this activity, you will get a chance to make mold grow. There are many types of mold, and if you are lucky, you will see several different kinds. You

The study of mold is a simple way to gain firsthand experience in experimenting and observing.

will be working with mold, but the real purpose of this activity is to give you some experience in experimenting and observing.

The format for this activity is slightly different from the other activities in that a structured reporting format is given to you, but you are on your own in devising your experiment. Some suggestions will be given, but you can also devise other experiments. You may need several weeks for this activity. The time factor will depend on how fast your mold develops. Allow at least two weeks from the time you first see mold. Don't forget that you can supplement this activity with some research in the library.

EQUIPMENT

Bread—the amount will depend on your experiments. Divide each slice into quarters for ease in handling. Commercial bread usually has preservatives which may slow down molding, but it will work with a little time and care. White bread usually works best because the colors show up better.

Container—Plastic bags are excellent. You can tie them closed and they are airtight. Small jars, such as baby food jars, also work well. You might find other containers that work just as well, but be sure that you can see both sides of your bread.

Water—*Important!* Bread needs to be moist (not wet) for best results. You might want to check this out as an experiment. Ten drops of water on a quarter of a slice of bread is recommended, then seal it to prevent evaporation. If the bread dries out, you will need to add more water.

Hand lens—A useful piece of equipment to help you see some interesting characteristics not usually seen with the naked eye. You might be able to learn some things about mold that you never knew before.

Other culture media—You might want to try growing mold on something besides bread, so you have a choice here. What will grow mold? How can you find out?

This teacher is preparing the pieces of bread for the mold experiment.

PROCEDURE

1. Cut a slice of bread into quarters. Put them on a clean surface and try not to touch them as you do this. Keep them as sterile as possible.
2. Decide how you want to treat each one. Keep one untreated for a control. Suggestions:
   a. Rub one quarter between your hands or on your cheek
   b. Rub one quarter on the floor or on the kitchen table.
   c. Rub one quarter on one of your textbooks.
   d. Your choice; use your imagination.
3. Put ten drops of water on each piece of bread.
4. Put each piece into a plastic bag and seal. *Attach a label to each bag describing the surface the bread was exposed to.*
5. Watch the samples and keep a record of what happens.
6. Make up an experiment of your own and try it.

RECORD KEEPING

   A. Experiment

   1. What was your variable?

2. What were your controls?

3. What problem are you trying to solve?

4. What do you think will happen?

B. Observations

Start your experiment and watch all of your samples daily. When you observe the first sign of mold on any of the samples, start recording the data for all of the samples. Data will be collected on the third, fifth, seventh, and tenth days after the first mold is observed.

1. How many days before you first saw mold?

   On which sample?

2. Third Day. Draw and color what you see.

COMMENTS

3. Fifth Day. Draw and color what you see.

BAG 1   BAG 2   BAG 3   BAG 4

How many colors do you see?

Which grows the fastest?

Which grows the slowest?

COMMENTS

4. Seventh Day. Draw and color what you see.

How many colors do you see?

Has anything significant happened between your last observation and this one? What was it?

COMMENTS

5. Tenth Day. Draw and color what you see.

Which color seemed to take over?

What do you think will happen five days from now?

COMMENTS

C. Supplementary Work

1. Have you looked at the mold through a hand lens? Draw a picture of what you saw and describe it.

2. Have you done any reading about mold? What did you read?

   What did you learn that interested you most?

3. What is one thing that you have learned about mold from your observations that you find most interesting?

4. Make any personal comments that you may have.

D. Experimental Variations

1. Design an experiment that you could do with mold. Describe it briefly here and develop it fully on a separate sheet. You need not actually do the experiment, *unless:* (a) You want to satisfy your curiosity, or (b) your instructor wants you to.
   Description:

   Problem:

   Variable:

   Controls:

    E. Use of the Activity in Teaching

1. How could you use this activity in an elementary science classroom?

2. List three references that you could use with your elementary students. Give title, author, publisher.

    a.

    b.

    c.

F. Present your results to your instructor for final check. *Don't throw your mold away until after this check.* Your instructor might want to see it as part of the check.

COMMENTS

Summary: Project Ideas

This concludes the activities for Part 6-2. You were given the opportunity to do some activities with living organisms. Each one used a different format so that you could learn different ways to present an activity to your students. Remember that children enjoy living organisms, large and small, and a great many things can be taught using this as a vehicle. In designing your own activities, think about the processes of science that you have used in these activities, as well as the subject matter content. Hopefully, you have had fun while you were learning and will construct activities for your students with this in mind.

FINAL SEMINAR

You may review your activities with your instructor or with your classmates. Try to relate the activities that you have completed to children and how they might react to the activities. Be prepared to discuss ways to use the activities in a classroom.

NOTES

COMPETENCY EVALUATION

You should have some knowledge of several activities that you can use with children. Your instructor will determine if a competency evaluation measure is necessary and, if so, what kind.

INDEX